Handbook of Wireless and Mobile Communications

Handbook of Wireless and Mobile Communications

Edited by **Benito Olin**

CWILLFORD PRESS

New York

Published by Willford Press,
118-35 Queens Blvd., Suite 400,
Forest Hills, NY 11375, USA
www.willfordpress.com

Handbook of Wireless and Mobile Communications
Edited by Benito Olin

International Standard Book Number: 978-1-68285-130-2 (Hardback)

Contents

               Protocol using Directional Antennas for MANETs   111
               Sandhya Chilukuri, Rinki Sharma, Deepali. R. Borade and
               Govind R. Kadambi

Chapter 10   **Performance of the IEEE 802.15.4a UWB System using Two**
               Pulse Shaping Techniques in Presence of Single and
               Double Narrowband Interferences   129
               Rasha S. El Khamy,.ShawkyShaaban, Ibrahim Ghaleb and
               Hassan Nadir Kheirallah

Chapter 11   **A New UWB System Based on a Frequency Domain**
               Transformation of the Received Signal   139
               Karima Ben Hamida El Abri and Ammar Bouallegue

Chapter 12   **Rach Congestion in Vehicular Networking**   153
               Ramprasad Subramanian and Kumbesan Sandrasegaran

Chapter 13   **A Low-Energy Fast Cyber for Aging Mechanism**
               for Mobile Devices   165
               Somayeh Kafaie, Omid Kashefi and Mohsen Sharifi

Chapter 14   **Distance Based Energy Efficient Selection of Nodes to**
               Cluster Head in Homogeneous Wireless Sensor Networks   177
               S. Taruna, Sheena Kohli and G. N. Purohit

Chapter 15   **Comparative Evaluation of Fading Channel Model**
               Selection for Mobile Wireless Transmission System   191
               Z. K. Adeyemo, D. O. Akande, F. K. Ojo and H. O. Raji

Chapter 16   **Parametric Performance Analysis of Patch Antenna using**
               EBG Substrate   202
               Mst. Nargis Aktar, Muhammad Shahin Uddin, Monir Morshed,
               Md. Ruhul Amin and Md. Mortuza Ali

               Permissions

               List of Contributors

Preface

Mobile communications are a primary sub-field of wireless communications. The past decade has witnessed rapid growth in both fields. This book traces the progress of these fields and highlights some of the key concepts and applications through thorough discussions of topics like wireless mesh networks, simulation, performance of mobile and wireless networks and systems, etc. This book will serve as a reference to a broad spectrum of readers and will be especially beneficial for students and researchers.

This book is a result of research of several months to collate the most relevant data in the field.

When I was approached with the idea of this book and the proposal to edit it, I was overwhelmed. It gave me an opportunity to reach out to all those who share a common interest with me in this field. I had 3 main parameters for editing this text:

1. Accuracy – The data and information provided in this book should be up-to-date and valuable to the readers.
2. Structure – The data must be presented in a structured format for easy understanding and better grasping of the readers.
3. Universal Approach – This book not only targets students but also experts and innovators in the field, thus my aim was to present topics which are of use to all.

Thus, it took me a couple of months to finish the editing of this book.

I would like to make a special mention of my publisher who considered me worthy of this opportunity and also supported me throughout the editing process. I would also like to thank the editing team at the back-end who extended their help whenever required.

Editor

MULTIUSER DETECTION IN ASYNCHRONOUS MULTIBEAM COMMUNICATIONS

Helmi Chaouech[1] and Ridha Bouallegue[2]

[1]National Engineering School of Tunis, University of El-Manar, Tunis, Tunisia
helmi.chaouech@planet.tn
[1,2]Innov'COM Laboratory, Sup'COM, University of Carthage, Tunis, Tunisia
ridha.bouallegue@gnet.tn

ABSTRACT

This paper deals with multi-user detection techniques in asynchronous multibeam satellite communications. The proposed solutions are based on successive interference cancellation architecture (SIC) and channel decoding algorithms. The aim of these detection methods is to reduce the effect of co-channel interference due to co-frequency access, and consequently, improves the capacity of the mulitbeam communications systems, by improving frequency reuse. Channel estimation allows the determination of interference coefficients, which helps their effects compensation. The developed multi-user detections techniques are iterative. Therefore, detection quality is improved from a stage to another. Moreover, a signals combining method, which is integrated into these detection solutions, enhances their capability. The proposed solutions are evaluated through computer simulations, where an asynchronous multibeam satellite link is considered over an AWGN channel. The obtained simulation results showed the robustness of these multi-user detection techniques.

KEYWORDS

Mulibeam System, Co-Channel Interference, SIC, Channel Decoding, Signals Combining.

1. INTRODUCTION

With the explosive evolution of information and communications technologies, mobility and flexibility of terminals become a necessity for several kinds of services. Wireless systems constitute a basic solution to satisfy these needs. In fact, they guarantee wireless and mobile communications for users independently of their localities. Thus, wireless communications solved some deficiencies of wired solutions. These contributions are mainly due to the transmission support; the atmosphere. The last allows data transmission every where; but an efficient physical resource sharing between users is necessary. Moreover, wireless transmissions are subject of several problems such as fading, noise, interference, multipath etc… These natural problems are dealt with robust signal processing techniques at the receivers.

As an example of wireless systems, satellite stations are good solutions which provide wide coverage and different communications services. These systems can guarantee network coverage in isolated places, where wired infrastructures and terrestrial wireless stations are difficult to install. Satellite systems offer also some specific services such as broadcasting, localisation, tracking etc.

Wireless channel is a common transmission support, which is shared between several simultaneous communications. Thus, efficient access and use of this communication mean is of great importance. Frequency is the main characteristic of wireless channels. Then, optimal division and reuse of this physical resource is needed to design high capacity communications systems. For wireless terrestrial networks, adjacent transceiver stations use different frequencies. In satellite systems, multibeam technology is a good solution for frequency reuse. This method forms at the satellite receiver a throng of narrow beams instead of a single wide beam. Each beam is defined by it carrier frequency. Thus, these different beams cover terrestrial zones with a co-frequency reuse for no adjacent cells. As a result, the capacity of the satellite system increases at the cost of co-channel interferences (CCI) and multiple access interference (MAI). The problems, which are due to these interferences and the noise, will be dealt by the receiver.

In this paper, we have developed some multi-user detection techniques for asynchronous multibeam systems. The aim of these solutions is to deal with co-channel interference, and provide an efficient frequency reuse. The proposed solutions are based on channel decoding, channel estimation, and interference cancellation, which operate in iterative processes. Due to asynchronous access of users to the system, propagation delay estimation task is needed as a first processed operation by the receiver. The developed techniques take advantage from the spatial diversity due to satellite antenna array. Thus, a signals combining solution, integrated in the multi-user detection methods, allows signal to noise ratio (SNR) improvement which leads to better detection quality.

The remainder of this paper is organized as follows. Section 2 presents related work. In section 3, the signal model for transmission and at the input of multi-user detection techniques is detailed. Section 4 shows the architecture of the system with its different functional blocks, and it expresses mathematically each block's operation. In section 5, simulations results are presented with some interpretations. At last, section 6 gives some conclusions from this work and presents propositions which can be subjects for the future works.

2. RELATED WORK

In this section, we present an overview of some related solutions. In fact, multi-user detection and channel estimation are research fields of big importance, especially for wireless communications. Thus, in the last three decades, many authors and several works are concentrated on detection and channel estimation techniques. Some of them, dealt with joint detection and channel estimation problems [1], [2], [3]. In fact, detection techniques need channel effects compensation in order to can combat noise and multiple access interference. In [4] and [5], some channel estimation techniques and propagation delay estimation methods are presented and evaluated. Some wireless channel models are discussed and evaluated in [6]. In [7], a channel estimation solution for mulibeam communication is developed. In [8], a channel estimation technique combined with a multi-user detection method is developed and evaluated. Evaluation of detection techniques can be done by bit error rate (BER) computation, and by their implementation complexity, as their processor time consummation. In his book [9], Verdu developed, analyzed and evaluated some multi-user detection techniques. Thus, based on their mathematical formulations, detection solutions can be classified into some categories. The conventional detection or the matched filtering detector suffers from co-channel interference and is very sensitive to near-far problem [9], [10]. The maximum likelihood (ML) detection technique, which is known as the optimal detector, is developed to solve optimally the weakness of the conventional detection [11]. This ML based detector has optimal performance in the presence of MAI and near-far problem at the cost of computational complexity, which does not promote its practical implementation particularly for real-time applications. The linear multi-user detectors, which are the decorrelating technique and the minimum mean-squared error (MMSE) detector, deal with MAI robustly, and they are near-far resistant [9], [10], [12], [13]. However, their major weaknesses are noise enhancement and inversion of big dimension matrices, especially in the case of asynchronous communications. Iterative interference cancellation detection techniques are robust solutions which subtract CCI and improve the BER iteratively [14], [15], [16], [17]. With these detection techniques, signals are cleaned of MAI successively or in a parallel processing. Thereafter, they can be classified in two main architectures; the successive interference cancellation (SIC) methods and the parallel interference cancellation (PIC) ones. Coded multi-user detections showed high performance. That's why; they occupy a big amount of the works which deal with detection problems. In fact, this kind of detection solutions, in addition to its interference cancellation, relies on channel decoders to combat noise and interference. These multi-user detection techniques give acceptable BERs even with low SNRs. Some detection techniques incorporating channel decoding are presented in [18], [19], [20], [21], [22]. Other detection solutions developed for multibeam communications, which are detailed and evaluated, can be found in [23], [24].

3. SIGNAL MODEL

We consider the asynchronous uplink of a multibeam satellite system. K active users share a co-frequency channel and send their signals to satellite antenna array. The K users belong to different cells which are covered by co-frequency beams. Users in the same cell adopt a TDMA access to physical resources. Thus, co-channel interferences are due to same frequency reuse. Mathematically, we represent the kth signal by:

$$r_k(t) = a_k e^{j\varphi_k} x_k(t) \tag{1}$$

Where, a_k and φ_k are the amplitude of the signal and its carrier phase, and $x_k(t)$ is given by:

$$x_k(t) = \sum_{i=0}^{N-1} x_k[i]g(t - iT - \tau_k) \tag{2}$$

With, $x_k[i]$, $i = 0..N-1$ is a sequence of N QPSK symbols, $g(t)$ is the emitter filter waveform, T is the symbol temporal duration and τ_k is the propagation delay of the kth signal. For simplicity and without loss of generality, we assume an ordering on the time delays such that: $\tau_1 \le \tau_2 \le \dots \le \tau_K < T$. We suppose that the propagation delays are multiple of T/N_s; the sampling period. Thus, N_s represents the sampling factor or the number of samples taken by symbol during. The sequence of N symbols, which are convolutional coded and interleaved before transmission, is divided into N_p pilot symbols and N_i information symbols.

This signal, expressed in (2), is received by the L radiating components of the antenna array, (see figure 1). Thus, if we generalize for the K users, the signal received by the lth element of the antenna can be expressed by:

$$s_l(t) = \sum_{k=1}^{K} d_k^{(l)} r_k(t) + n_l(t) \tag{3}$$

With, $d_k^{(l)}$ is the lth coefficient of the steering vector d_k of the kth received signal, and $n_l(t)$ is a Gaussian noise, added to the composite signal received by the lth antenna component.

To arrange the L signals received by the antenna array sensors, we define the signal vector $s(t)$ by: $s(t) = [s_1(t), s_2(t), \dots, s_L(t)]^T$. Where, $(.)^T$ denotes the transpose operator. Thus, generalization of equation (3) for the L components of the antenna gives:

$$s(t) = \sum_{k=1}^{K} d_k r_k(t) + n(t) \tag{4}$$

Where, d_k, as it is mentioned above, is the column vector of length L which contains information about the arrival direction of the kth signal [25], and $n(t) = [n_1(t), \ldots, n_L(t)]^T$ is the additive noise vector at the L radiating elements outputs. We define the direction of arrival (DOA) matrix of the K beams by: $D = [d_1, \ldots, d_K]$, and the vector of received signals: $r(t) = [r_1(t), \ldots, r_K(t)]^T$. Thus, equation (4) can be rewritten as follows:

$$s(t) = Dr(t) + n(t) \qquad (5)$$

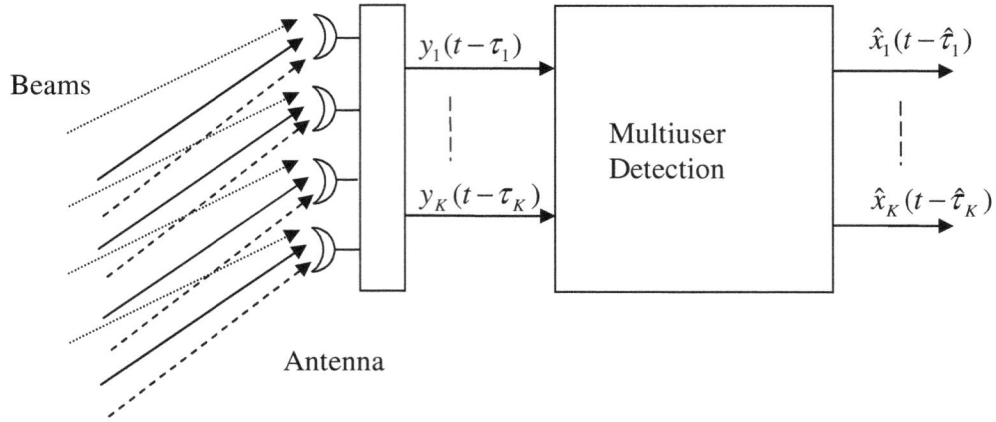

Figure 1. Reception model.

At the receiver, beam forming operation maximizes the energy of the useful signal by steering the antenna array in the DOA of this beam [25]. Thereafter, to form the kth beam, the treatment consists of taking a linear combination of the signals at the antenna elements outputs. The obtained signal for the kth beam is given by:

$$y_k(t) = v_k{}^T s(t) \qquad (6)$$

With, v_k is a column vector which contains the L coefficients of the kth beam forming. Generalization of the expression (6) for the K active users in the system gives:

$$y(t) = Vs(t) \qquad (7)$$

With, $y(t) = [y_1(t), y_2(t), \ldots, y_K(t)]^T$ and $V = [v_1, v_2, \ldots, v_K]^T$.

Using (5), equation (7) becomes:

$$y(t) = VDr(t) + Vn(t) = Wr(t) + Vn(t) \qquad (8)$$

We define the diagonal matrix of the K signals complex amplitudes by: $A = diag\left([a_1 e^{j\varphi_1}, \ldots, a_K e^{j\varphi_K}]\right)$. With $diag(.)$ denotes the matrix diagonal operator. Using the expression of $r(t)$ in (1), equation (8) can be rewritten also:

$$y(t) = WAx(t) + Vn(t) \qquad (9)$$

Where, $x(t) = [x_1(t), \ldots, x_K(t)]^T$

By introducing the channel matrix $H = WA$, and the noise vector $z(t) = Vn(t)$, equation (9) becomes:

$$y(t) = Hx(t) + z(t) \qquad (10)$$

The discreet model, which is derived from (10) after optimal sampling, is given by:

$$y[i] = Hx[i] + z[i] \qquad (11)$$

Thus, these samples will be dealt by the multi-user detection technique in order to determine the original data sent by user's terminals.

4. MULTI-USER DETECTION

The proposed multi-user detection techniques are composed of some functional blocks which operate the following jobs: propagation delay estimation, phase estimation, channel decoding, channel estimation, interference cancellation and signals combining. These operations are executed successively. In each block of the receiver, signals users are also dealt successively. Some of these operations are processed iteratively in some stages and the others are dealt only at the beginning of algorithms running.

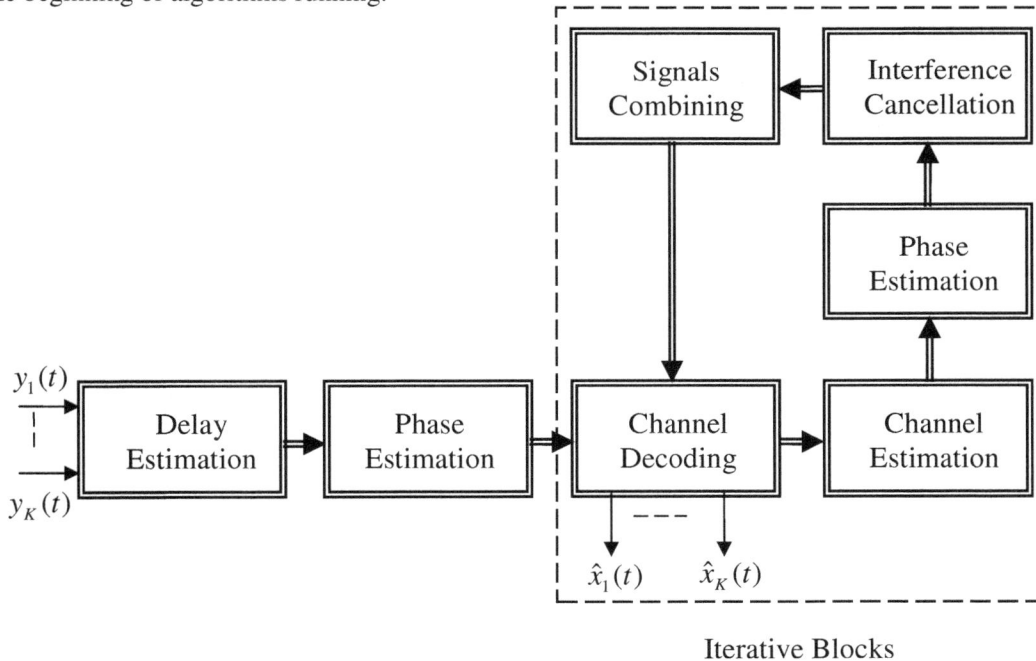

Figure 2. Receiver architecture.

4.1. Delay Estimation

The first operation to be dealt is the propagation delays estimation. The results of this operation are then used by the following receiver blocks. Thus, the performance of the multi-user detection is strongly influenced by the delays estimation performances. In each processing operation of the receiver, the asynchronous character of the signals is not modified but taken into account. The propagation delay estimation technique is based on signals correlation, and it uses the pilot symbols. For the kth signal, estimation of its propagation delay can be expressed by:

$$\hat{\tau}_k = \arg Max\left(\left|\sum_{i=1}^{N_s \times N_p} \mathrm{Re}\big(y_k(i).P_k(i)^*\big)\right| + \left|\sum_{i=1}^{N_s \times N_p} \mathrm{Im}\big(y_k(i).P_k(i)^*\big)\right|\right) \tag{12}$$

Where, $\mathrm{Re}(.)$ and $\mathrm{Im}(.)$ denote the real part and imaginary part of a complex number respectively, and $(.)^*$ denotes the conjugate operator of complex number. N_p is the pilot symbols number.

4.2. Phase Estimation

From figure 3, we note that the receiver contains two phase estimation blocks. In fact, these two operations are done differently. The first is an initialisation of the iterative algorithm. It is needed to compensate the phase effects before decoding algorithms processing. In the first stage of the multi-user detection techniques, an initial phase estimation of the kth signal is given by:

$$\hat{\varphi}_k = ang\left(\sum_{N_p} y_k(j)\tilde{P}_k(j)^*\right) \tag{13}$$

Where, $\tilde{P}_k$ is a $N_s \times N_p$ length column vector. It is derived from the pilot sequence vector P_k as follows: $\tilde{P}_k = \big[P_k(1),..,P_k(1),\ldots,P_k(N_p),..,P_k(N_p)\big]^T$, with, P_k is the vector of kth user training sequence which consists of N_p pilot symbols. And, $ang(.)$ denotes the angle of a complex number operator.

The second phase estimation block, which is included in the iterative part of the algorithms, is performed after channel coefficients estimation. This iterative operation can be expressed, for the kth signal at the nth stage, as:

$$\hat{\varphi}_k^{(n)} = ang\left(\hat{h}_{k,k}^{(n)}\right) \tag{14}$$

With, $\hat{h}_{k,k}^{(n)}$ is the (k, k)th channel matrix estimation at the nth iteration. It is explained in section 4.4.

4.3. Channel Decoding

Before channel decoding, the signals phases are compensated with use of phases estimations. This operation consists to multiply the symbols samples by the quantities $e^{-j\hat{\varphi}_k^{(n)}}$ for the K users respectively. We have implemented two different decoding methods for the convolutional channel coding. They are the Viterbi algorithm [26], [27], and the BCJR one [28], [29]. The second technique needs SNR knowledge. Thus, SNR estimation operation, which is not presented in the above receiver architecture, is necessary. This task is explained in the following paragraph.

4.3.1. SNR estimation

The SNR estimation is computed in each iteration before BCJR or MAP (Maximum a Posteriori) decoding operation. It is based on the BCJR algorithm itself. That solution provides an estimation of the signal to noise ratio by minimizing the bit error rate between the pilot symbols and the decoded samples which correspond to the transmission of that pilot sequence. Thus, SNR estimation can be described by:

$$\hat{SNR} = \arg\left(\min_{snr \in [snr_{min} \ snr_{max}]} (BER) \right) \tag{15}$$

Other solutions of SNR estimation techniques can be found in [30].

4.4. Channel Estimation

The channel estimation technique allows the determination of CCI coefficients in order to compensate their effect in interference cancellation block. These coefficients, which are the channel matrix H elements, are estimated iteratively. In each stage, they are updated, with use of estimated and pilot symbols. Channel coefficients estimation of the kth signal at the nth iteration can be expressed by the following equation:

$$\hat{h}_{k,l}^{(n)} = \frac{1}{2 \times N_s(N+1)} \sum_{j=1}^{N_s(N+1)} y_k(j)\hat{x}_l^{(n)}(j) \tag{16}$$

Where,

- $\hat{h}_{k,l}^{(n)}$ is the estimation of interference coefficient of signal l on signal k at n^{th} iteration, or the $(k,l)^{th}$ element estimation of the matrix H at the at the nth iteration.
- The value 2 in the denominator is to compensate the effect of the square of the QPSK symbols modulus on the channel coefficients estimation.
- $\hat{x}_l^{(n)}(j)$ is the estimation of the jth symbol of user l at the nth iteration.

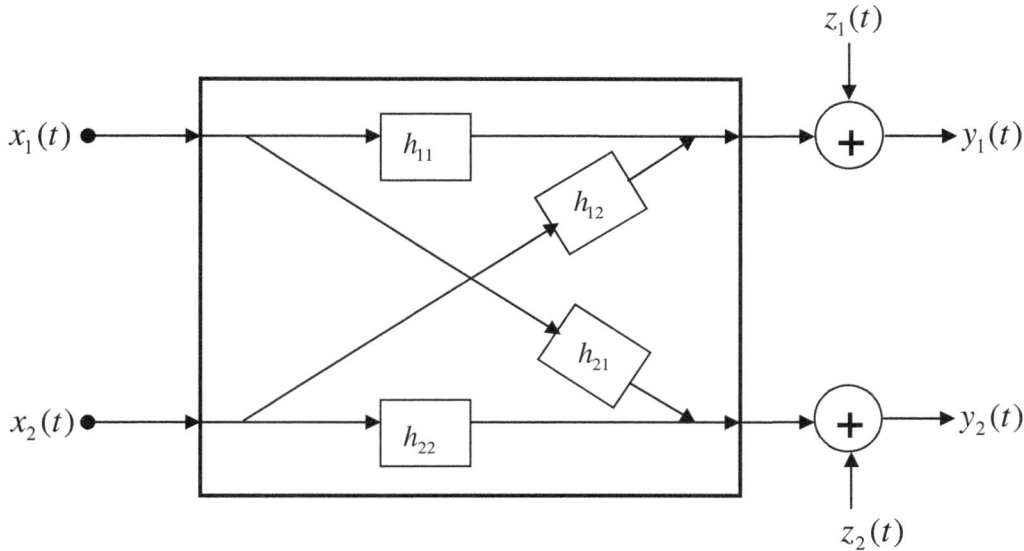

Figure 3. CCI Channel Model for K=2.

4.5. Interference Cancellation

The successive interference cancellation block ensures the CCI suppression from the K beams. Thus, co-frequency signals are dealt successively in each iteration. Symbols of interfering signals are extracted from the useful signal with use of the appropriate estimated channel coefficients, and taking into account of signals propagation delays. Thus, interference cancellation from the jth sample of kth signal at the nth iteration is expressed by:

$$y_k^{(n)}(j) = y_k(j) - \sum_{k'=1}^{k-1} \hat{h}_k^{(n-1)}(k') \, x_{k'}^{(n)}(j) - \sum_{k'=k+1}^{K} \hat{h}_k^{(n-1)}(k') \, x_{k'}^{(n-1)}(j) \qquad (17)$$

4.6. Signals Combining

The technique of signals combining takes advantage from the spatial diversity due to the antenna array. This solution improves the SNR of the useful signal by extracting its interfering parts on the other beams, and adds them coherently to this signal. In this way, that method leads to better detection quality. Thus, for the jth sample of kth signal at the nth stage of the multi-user detection techniques, signals combining operation can be expressed by the two following equations:

$$\chi_k^{(n)}(j) = \sum_{\substack{k'=1 \\ k' \neq k}}^{K} \hat{h}_{k'}^{(n-1)}(k)^* \left(y_{k'}(j) - \sum_{\substack{k''=1 \\ k'' \neq k'}}^{K} \hat{h}_{k'}^{(n-1)}(k'') \hat{x}_{k''}^{(n-1)}(j) \right) \qquad (18)$$

$$\tilde{y}_k^{(n)}(j) = y_k^{(n)}(j) + \chi_k^{(n)}(j) \qquad (19)$$

Thereafter, at the nth iteration, for the jth symbol of the kth signal, the decoder computes the quantity $\tilde{y}_k^{(n)}(j)$.

5. SIMULATIONS RESULTS

In order to evaluate the performances of the multi-user detection techniques, we have considered an asynchronous multibeam satellite reverse link scenario. Channel coding and modulation are taken those of the DVB RCS standard [31]. Thus, we have employed the convolutional coding of this system to code the transmission data frame. This convolutional code is defined by its two generator polynomials, which are represented by their octal form as: [171 133]. The modulation is QPSK. Before transmission, data is randomly interleaved. The other parameters of simulation, which characterize the applied scenario, are: K=5, Ni=100, Np=30 and Ns=4. We have performed the multi-user detection techniques with Monte Carlo simulations. In each iteration, N QPSK symbols and K carrier phases are randomly generated. The phases are uniformly distributed in $[0, 2\pi]$. The propagation delays, which belong to the temporal interval $[0, T[$, are also randomly generated. The signals amplitudes are taken equal, with unit powers; $a_1 = a_2 = \ldots = a_5 = 1$. We considered that the interferences are uniformly distributed between the co-frequency signals. Then, the CCI coefficients are of the same modulus. Thus, the CCI channel matrix can be presented by:

$$|H| = \begin{bmatrix} 1 & \mu & \mu & \mu & \mu \\ \mu & 1 & \mu & \mu & \mu \\ \mu & \mu & 1 & \mu & \mu \\ \mu & \mu & \mu & 1 & \mu \\ \mu & \mu & \mu & \mu & 1 \end{bmatrix} \tag{20}$$

Where, $\mu = 0.25$.

The following figures show the simulations results. In order to evaluate the performances of the multiuser detection techniques, we have plotted in figures 4 and 5 the average BER evolution with different values of SNRs. The propagation delays estimation method is evaluated in figure 6, by computing delay estimation errors vs SNRs. In figure 4, the simulations results of the multi-user detection technique with Viterbi channel decoding are presented. The performances of this solution are good even with low SNRs. as an iterative algorithm, the detection is improved from a stage to another of the SIC multi-user architecture. We have shown, in figure 5, the simulation results of the detection technique implementing BCRJ channel decoding. Compared to the first solution, the obtained results, with MAP decoding, are better. But, when we ran algorithms via computer simulations, we noted that the viterbi decoding based technique is more than three once faster than the solution implementing BCJR algorithm. Thus, MAP decoding introduced, with the significant improvement performances, some complexity of processing at the receiver. Moreover, SNR estimation operation, which is performed before BCJR, decoding, consumed some processing time. If we look again on simulations results shown in figures 4 and 5, it is clear that signals combining technique enhanced the detection quality for both solutions. This operation allowed somehow the decreasing of stages number of the algorithms by converging rapidly to the desired bit error rates. For example, according to figures 4 and 5, stages 2, where signals combining techniques are applied, gave better results than stages 3 without signals combining application. In figure 6, we have evaluated the propagation delays estimation technique. Average error of delays estimation of the K users is presented for different values of the SNRs and CCI powers. The obtained results show the robustness of the proposed solution under a highly noisy environment, although some degradation of estimation quality, which is due to co-channel interference power increasing. Thus, high estimation quality of the propagation delays was of great importance for the other detection blocks, since the signals are dealt with asynchronously, without need of their synchronization as a prior task.

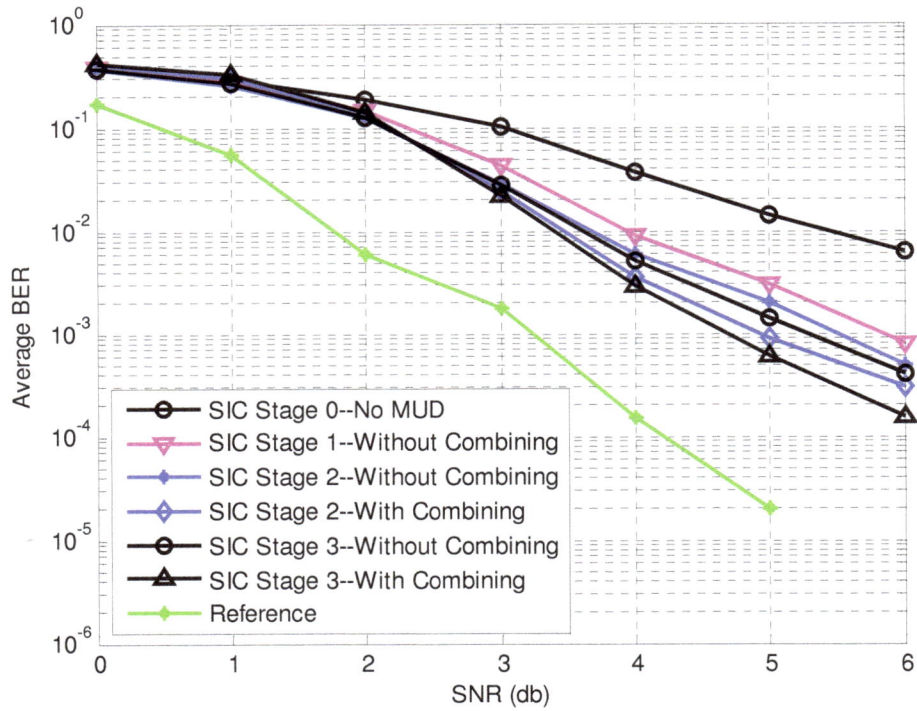

Figure 4. Performance evaluation of the detection with Viterbi decoding.

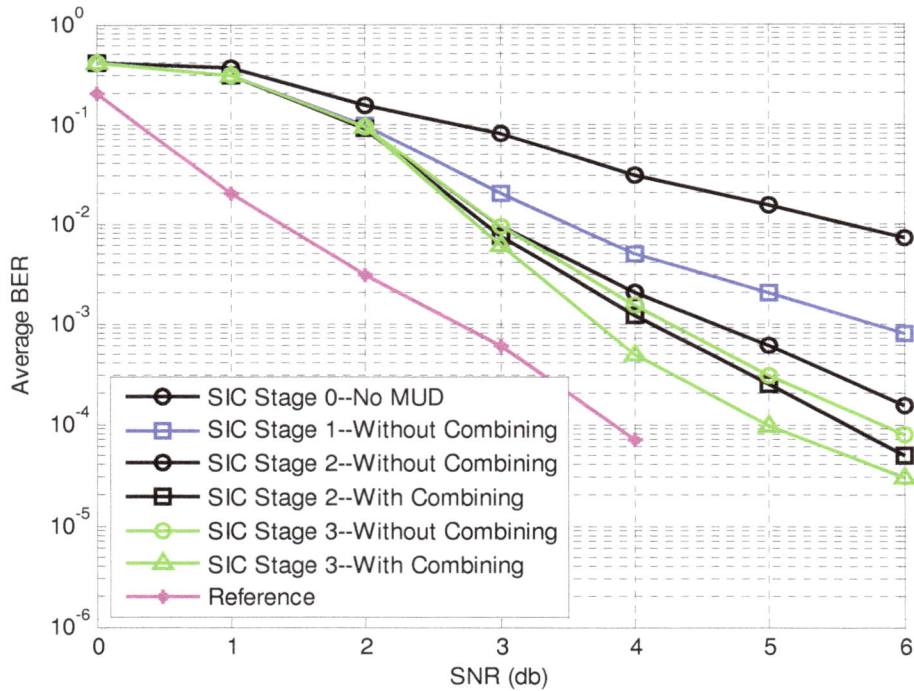

Figure 5. Performance evaluation of the detection with BCJR decoding.

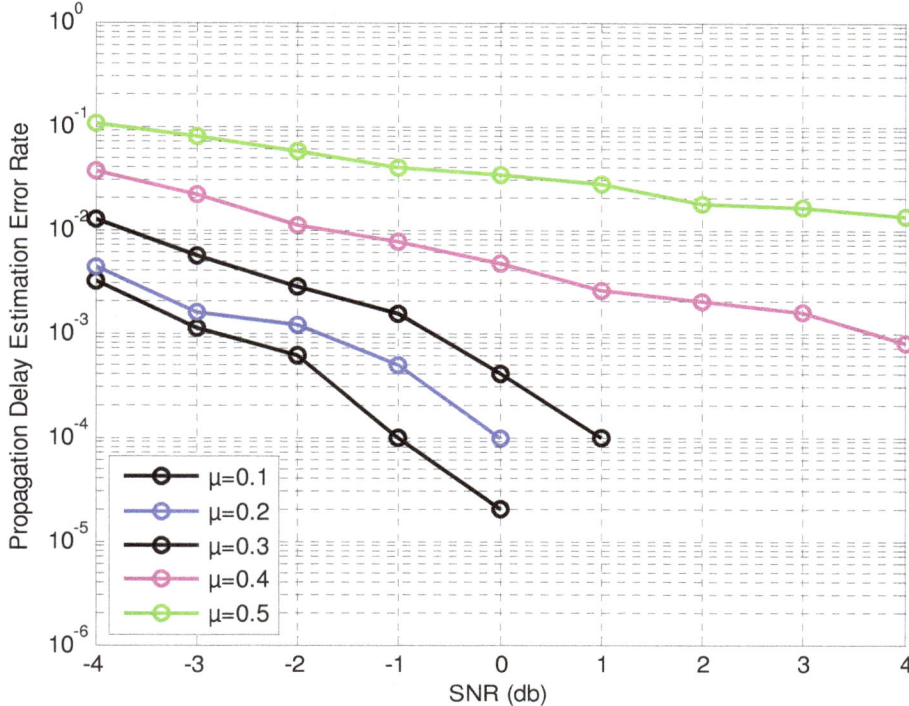

Figure 6. Performance evaluation of the propagation delay estimation.

6. CONCLUSIONS

In this paper, we have developed and evaluated through computer simulations some multi-user detection solutions for asynchronous multibeam systems. The proposed techniques are based on successive interference cancellation architectures which implement channel decoding and estimation methods. We integrated two convolutional decoding algorithms. The Viterbi technique and the BCJR one. Both multi-user detection techniques showed good performances under noisy and CCI situation. Moreover, the quality of symbols detection is improved from a stage to another. The solution, which implements MAP decoding, gave better results compared to the detection technique employing Viterbi decoding algorithm. But, it introduced more processing complexity in the receiver due to probabilities calculation, in addition to the SNR estimation block, which is needed for MAP decoding. Thus, the multi-user solution, which integrates Viterbi algorithm, is faster in execution, than the other. As we dealt with asynchronous communications, estimation of propagation delays operation was of significant importance. Its good performances helped enormously detection techniques processing. Through the simulations results, we can conclude the importance of the signals combining technique. This operation allowed SNR improvement of the useful signal, which leads to a better quality of detection, and therefore, a possible reduction of the stages number. As a future work, it will be interesting to deal with multi-user detection techniques in multibeam communications under an asynchronous multipath channel.

REFERENCES

[1]   K. J. Kim and R.A. Iltis, "Joint detection and channel estimation algorithms for QS-CDMA signals over time-varying channels," IEEE Transactions on communications, Vol. 50, No. 5, May 2002.

[2]   T. Zemen, C. F. Mecklenbrauker, J. Wehenger and R.R. Muller, "Iterative joint time-variant channel estimation and multi-user detection for MC-CDMA," IEEE Transactions on wireless communications, Vol. 5, No. 6, June 2006.

[3]   S. Y. Park, B. Seo and C. Kang, "Performance of iterative receiver for joint detection and channel estimation in SDM/OFDM systems," IEICE Transactions on communications, Vol. E86-B, No. 3, March 2003.

[4]   M. Sirbu, "Channel and delay estimation algorithms for wireless communication systems," Thesis of Helsinki University of Technology (Espoo, Finland), December 2003.

[5]   L. M. Davis, I. B. Collings and R. J. Evans, " Estimation of LEO satellite channels," International conference on information, communications and signal processing, pp. 15-19, Singapore, September 1997.

[6]   A. Amer and F. Gebali, "General model for infrastructure multichannel wireless LANs," International Journal of Computer Networks & Sommunications (IJCNC), Vol. 2, No. 3, May 2010.

[7]   H. Chaouech and R. Bouallegue, "Channel estimation and detection for multibeam satellite communications," IEEE Asia pacific conference on circuits and systems, pp. 366-369, Kuala Lumpur, Malaysia, December 2010.

[8]   H. Chaouech and R. Bouallegue, "Channel estimation and multiuser detection in asynchronous satellite communications," International Journal of Wireless & Mobile Networks, Vol. 2, No. 4, pp. 126-139, November 2010.

[9]   S. Verdu, Multiuser Detection, Cambridge University Press, 1998.

[10]   K. Khairnar and S. Nema, "Comparison of Multi-user detectors of DS-CDMA System," World Academy of Science, Engineering and Technology, 2005.

[11]   S. Verdu, "Minimum probability of error for asynchronous gaussian multiple access channels," IEEE Transactions on information theory, Vol. IT-32, No. 1, January 1986.

[12]   R. Lupas and S. Verdu, "Linear multi-user detectors for synchronous code-division multiple access channels," IEEE Transactions on information theory, Vol. 35, No. 1, pp. 123-136, January 1989.

[13]   R. Lupas and S. Verdu, "Near-far resistance of multiuser detectors in asynchronous channels," IEEE Transactions on communications, Vol. 38, No. 4, pp. 496-508, April 1990.

[14]   M. K. Varanasi, "Multistage detection in asynchronous code-division multiple-access communications," IEEE Transactions on communications, Vol. 38, No. 4, pp. 509-519, April 1990.

[14]   A. L. C. Hui and K. Ben Letaief, "Successive interference cancellation for multi-user asynchronous DS/CDMA detectors in multipath fading links," IEEE Transactions on communications, Vol. 46, No. 3, pp. 384-391, March 1998.

[16]   K. Ko, M. Joo, H. Lee, and D. Hong, "Performance analysis for multistage interference cancellers in asynchronous DS-CDMA Systems," IEEE Communications letters, Vol. 6, No. 12, December 2002.

[17]   S. H. Han and J. H. Lee, "Multi-stage partial parallel interference cancellation receivers for multi-rate DS-CDMA system," IEICE Transactions on communications, Vol. E86-B, No. 1, January 2003.

[18]   J. Hagenauer, E. Offer, and L. Papke, "Iterative decoding of binary block and convolutional codes," IEEE Transactions on information theory, Vol. 42, No. 2, pp.429-445, March 1996.

[19]   M. L. Moher, "An iterative multiuser decoder for near-capacity communications," IEEE Transactions on communications, Vol. 46, No. 7, pp. 970-880, July 1998.

[20]   M. L. Moher, "An iterative algorithm for asynchronous coded multi-user detection," IEEE Communications letters, Vol. 2, No. 8, pp. 229-231, August 1998.

[21]   H. H. Chen and Z. Q. Liu, "A CDMA multi-user detector with block channel coding and its performance analysis under multiple access interference," IEICE Transactions on communications letters, Vol. E81-B, No. 5, pp. 1095-1101, May 1998.

[22]   C. Schlegel, P. Alexander and S. Roy, "Coded asynchronous CDMA and its efficient detection," IEEE Transactions on information theory, Vol. 44, No. 7, pp. 2837-2847, November 1998.

[23]   M. L. Moher, "Multiuser decoding for multibeam systems," IEEE Transactions on vehicular technology, Vol. 49, No. 4, pp. 1226-1234, July 2000.

[24]   M. Debbah, G. Gallinaro, R. Muller, R. Rinaldo, and A. Vernucci, "Interference mitigation for the reverse- link of interactive satellite networks," 9th International workshop on signal processing for space communications (SPSC), ESTEC, Noordwijk, The Netherlands, 11-13 September 2006.

[25]   Lal. C. Godara, "Application of antenna arrays to mobile communications, Part II: Beam-forming and direction-of-arrival considerations," Proceedings of the IEEE, Vol. 85, No. 8, August 1997.

[26]   A. J. Viterbi, "Error bounds for convolutional codes and an asymptotically optimum decoding algorithm," IEEE Transactions on information theory, Vol. 13, pp. 260-269, April 1967.

[27]   G. D. Forney, "The Viterbi algorithm," Proceedings of the IEEE, Vol. 61, No. 3, pp.268-278, March 1973.

[28]   L. R. Bahl, J. Cocke, F. Jelinek, and J. Raviv, "Optimal decoding of linear codes for minimizing symbol error rate," IEEE Transactions on information theory, Vol. IT-20, pp. 284-287, March 1974.

[29]   P. Robertson, P. Villebrun, and P. Hoeher, "A comparison of optimal and sub-optimal MAP decoding algorithms operating in the log domain," IEEE International conference on communications, Seattle, Washington, June 1995, pp. 1009-1013.

[30]   D. R. Pauluzzi and N. C. Beaulieu, "A comparison of SNR estimation techniques for the AWGN channel," IEEE Transactions on communications, Vol. 48, pp. 1681-1691, October 2000.

[31]   ETSI EN 301 790 V1.3.1 (2003-03), "Digital Video Brodcasting (DVB); interaction channel for satellite distribution systems".

2

On the Use of Smart Ants for Efficient Routing in Wireless Mesh Networks

Fawaz Bokhari and Gergely Zaruba

Department of Computer Science and Engineering, The University of Texas at Arlington, Texas, US
fawaz.bokhari@mavs.uta.edu, zaruba@cse.uta.edu

ABSTRACT

Routing in wireless mesh networks (WMNs) has been an active area of research for the last several years. In this paper, we address the problem of packet routing for efficient data forwarding in wireless mesh networks (WMNs) with the help of smart ants acting as intelligent agents. The aim of this paper is to study the use of such biologically inspired agents to effectively route the packets in WMNs. In particular, we propose AntMesh, a distributed interference-aware data forwarding algorithm which enables the use of smart ants to probabilistically and concurrently perform the routing and data forwarding in order to stochastically solve a dynamic network routing problem. AntMesh belongs to the class of routing algorithms inspired by the behaviour of real ants which are known to find a shortest path between their nest and a food source. In addition, AntMesh has the capability to effectively utilize the space/channel diversity typically common in multi radio WMNs and to discover high throughput paths with less inter-flow and intra-flow interference while conventional wireless network routing protocols fail to do so. We implement our smart ant-based routing algorithm in ns-2 and carry out extensive evaluation. We demonstrate the stability of AntMesh in terms of how quickly it adapts itself to the changing dynamics or load on the network. We tune the parameters of AntMesh algorithm to study the effect on its performance in terms of the routing load and end-to-end delay and have tested its performance under various network scenarios particularly fixed nodes mesh networks and also on mobile WMN scenarios. The results obtained show AntMesh's advantages that make it a valuable candidate to operate in mesh networks.

KEYWORDS

Wireless mesh networks, link interference, ant colony optimization, routing, meta-heuristic.

1. INTRODUCTION

Wireless mesh networks (WMNs) have emerged as a promising technology for the provision of last mile broadband Internet access infrastructure in both urban and rural environments. Such networks are characterized as fixed backbone WMNs where relay nodes are generally static and are mostly supplied by permanent power sources [1]. With the availability of off-the-shelf, low cost, commodity networking hardware, it is possible to incorporate multiple radio interfaces operating in different radio channels on a single mesh router; thus forming a multi-radio mesh network. This enables a potential large improvement in the capacity of mesh networks compared to single-radio mesh networks [20]. However, due to the shared nature of the wireless medium, nodes compete with each other to access the wireless channel, resulting in possible interference among nodes which may affect traffic in the network. Furthermore, the fact that nodal density in a typical WMN is high compared to other wireless networks, make the flow interference issue in WMNs more severe. Incorporating techniques to specifically address these characteristics in a WMN routing protocol could improve the overall network flow capacity or performance of individual flows in the network.

Generally, routing algorithms can be defined as multi-objective optimization problems in a dynamic stochastic environment. However, formalizing routing as such optimization problem requires complete knowledge of traffic flows between each node in the network; this is prohibitively difficult to model in the presence of rapidly changing network dynamics (found in typical WMNs). Therefore, heuristic policies are often used to create quasi-optimal routing in WMNs. There has been a significant body of research in designing efficient heuristic based routing protocols and metrics for WMNs (for a quick overview see [2 - 5]). In the wired networking domains, a new family of routing algorithms has been proposed based on swarm intelligence by Dorigo et al. called Ant Colony Optimization (ACO) framework [6, 7], which is a meta-heuristic approach for solving hard optimization problems.

ACO algorithms draw their inspiration from the behaviour of real ants, which are known to find the shortest path between their nest and a food source by a process where they deposit pheromones along trails (acting like a local message exchange in a communication network). Ants generally start out moving at random, however, when they encounter a previously laid trail, they can decide to follow it, thus reinforcing the trail with their own pheromone substance. This process is thus characterized as a positive feedback loop, where the probability with which an ant chooses a path increases with the number of ants that previously chose the same path [7]. Hence, In ACO framework, *artificial ants* probabilistically and iteratively converge to a solution by taking into account pheromone trails deposited by other ants and other local heuristics. Artificial ants move stochastically (instead of deterministically) in the solution space, therefore they can explore a wider variety of possible solutions of a problem independently and in parallel. A more detailed explanation of the ACO framework can be found in [6].

1.1 Problem Addressed

In this paper, we address the problem of packet routing for efficient data forwarding in wireless mesh networks. Since most of the traffic in a mesh network usually flows between regular nodes and a few Internet gateways (i.e., rarely end-to-end between regular nodes). This can result in an uneven loading of links and can cause certain paths to be saturated. Similarly, the existence of inter-flow interference among the nodes and intra-flow interference within a transmission path may affect traffic loads on mesh nodes in a multi-radio WMN. The main objective of any mesh routing protocol is thus to effectively distribute the traffic by selecting channel diverse paths with less inter/intra flow interference.

1.2 Our Contributions

The salient features of our work that set us apart from the existing routing protocols in WMNs are listed as follows:

- We propose a distributed routing mechanism which enables the use of smart ants to probabilistically and concurrently perform the routing and data forwarding in order to stochastically solve a dynamic network routing problem.
- We formally define the properties and conditions necessary in the design of such smart ant based routing algorithm for WMNs. These smart ants help to effectively utilize the space/channel diversity typically common in multi radio WMNs.
- One interesting result is that smart ants has the capability to discover high throughput paths with less inter and intra-flow interference when conventional wireless network routing protocols fail to do so.
- The stability of any routing protocol depends upon how quickly it adapts itself to the changing dynamics of the network. We demonstrate through simulations that AntMesh quickly converges to the best path under situations when traffic characteristics change. We tune the parameters of AntMesh algorithm to study the effect of these on the performance in terms of the routing load and end-to-end delay.
- To the best of our knowledge, our proposed forwarding technique is among the first works that investigates the use of smart ants in WMNs and demonstrates a possible good performance advantage.

1.3 Paper Organization

The rest of the paper is organized as follows. We start with providing an overview of some of the existing ACO based routing protocols in wireless networks in Section II. The concept of smart ants in mesh networks and the necessary properties that they should possess are discussed in Section III. We will explain the working of our smart ant-based routing algorithm (AntMesh) in Section IV together with the description of how smart ants are used in AntMesh to capture the traffic load and inter/intra flow interference in WMNs. We describe the implementation details of AntMesh followed by simulation results to evaluate its performance to existing routing schemes in Section V. Finally we conclude the paper in Section VI.

2. RELATED WORK

There have been research efforts in using ant-based techniques for efficient data forwarding in wired networks [8, 9, 13] and mobile ad-hoc networks (MANETs) [10-12, 25-28, 34, 35]. This section provides an overview of some of the more prevalent ant-inspired routing algorithms in communication networks.

AntNet: one of the first applications of ant colony optimization framework (ACO) for wired network routing is AntNet [8]. In AntNet, routing is achieved by generating forward ants at regular intervals from a source node to a destination node to discover a low cost path and by backward ants that travel from destination to source to update pheromone table at each intermediate node. Forward ants keep track of trip times from source to destination node using the data traffic queues in order to experience the same delay that data packets experience. Forward ants select the next hop by a probabilistic decision rule which takes into consideration the pheromone intensity which is reinforced by other backward ants and heuristic information which is based on the queue length of the intermediate node. Once, a forward ant reaches the destination node, a backward ant is generated that tracks back to the source using high priority queues (for timely delivery) reinforcing the selection probability of intermediate nodes according to the fitness of the trip times of forward ants. However, the fact that AntNet was proposed for wired networks makes it unsuitable for WMNs because of their unique characteristics i.e. wireless interference and load balancing etc. etc. There are variations and extensions, e.g., [13, 14], of the original AntNet algorithm which targeted wired networks and thus are outside the scope of our research.

AntHocNet [12]: is a hybrid, multi-path, ant based algorithm. It consists of both a reactive and a proactive component. The reactive part is used for route establishment whereas the proactive component is used for route maintenance. The reactive component is used at the start of a data session where a *reactive forward ant* is broadcast to find multiple paths to the destination and upon reaching the destination; a *reactive backward ant* sets up the multiple paths to the destination using local heuristic information. While data packets are being routed, proactive forward ants are also generated. This helps in exploring new paths and getting up-to-date link quality information. One of the drawbacks of AntHocNet is the number of ants that need to be sent over the network for establishing routes to destinations as they are broadcast during a route discovery phase. Also, each ant stores list of visited nodes from source to current node and depending upon the distance to the destination, this list (and thus the ant's size) can grow long, increasing routing overhead.

POSANT: the authors of [10] present a position based ant colony routing algorithm for MANETs, in which they make use of position based instruments (e.g., GPS receivers) to combine with the ACO technique. POSANT is a reactive routing algorithm and [10] argues that the use of position information can greatly reduce the number of ant generations while reducing the route establishment time as well. Position information is used in the heuristic maintained at each node helping ants decide what next hop to take in the path discovery phase. However,

POSANT makes an assumption that the transmission time of an ant to its neighbours is the same for all nodes therefore ignoring packet loss and flow interference, which are quite common in wireless networks.

SARA: Rosati *et al.*, [34] have proposed a distributed ant routing algorithm for ad hoc networks where nodes are frequently joining and leaving the network and minimum signalling overhead is required. One of the design objectives of their proposed approach was to have low computational complexity and for this particular purpose, the algorithm creates routes on-demand making it a reactive protocol. Moreover, ants in SARA algorithm store only the node identity information to avoid extra control overhead and only the pheromone value is taken to make a forwarding decision. The authors have conducted extensive simulations and showed the effectiveness of their algorithm in ad hoc networks with critical connectivity when compared with AODV. Since the approach aims at minimizing complexity in the nodes making it as simple as possible, the authors claim that SARA can be helpful in achieving seamless routing in heterogeneous networks. However, the same simplicity comes with the drawback that it does not capture the network dynamics in terms of link characteristics and interference and taking necessary steps accordingly. Therefore it is not suitable for networks with high nodal densities and traffic flows moving to mostly one side of the network (gateways) typically like in wireless mesh networks.

DAR: recently, the authors of [35] have presented a simple ant routing algorithm for ad hoc networks with the objective of minimizing the protocol overhead. To achieve this objective, they have proposed a control neighbour broadcast method in which, a broadcast ant is unicast to the second-hop neighbour. Similarly, data packets are used to refresh paths of the active sessions in order to reduce the control overhead. Their third approach is implemented in link failure situation to repair that link locally using the two ends of the link rather than initiating a new source to destination path request. They have conducted extensive simulations to prove the main objective of their proposed approach i.e. less signalling overhead when compared with other existing ant-based routing algorithms.

For wireless mesh networks, we presented an interference-aware routing scheme based on ACO framework for single and multi radio WMNs [15, 29]. With respect to the previously quoted works on AntMesh, this paper presents a deeper comparison analysis and insight on the performance of smart ants in WMNs. We study in depth the working of AntMesh routing algorithm and describe the desirable properties of smart ants particularly suited for multi radio WMNs. We demonstrate that the network performance can be optimized when AntMesh algorithm parameters are carefully selected. The use of smart ants for routing on mesh networks shows a noticeable advantage against some well-known existing biological routing approaches.

3. SMART ANTS IN MESH NETWORKS

This section explains the rational on the use of ants for routing in WMNs. We will explain the concept of the smart ants, how they differ from regular ants used in previous techniques and the necessary properties that these smart ants should possess in order to efficiently perform routing in WMNs.

3.1 Desirable Properties of Smart Ants

Let us consider what kind of ants would be suitable for creating network paths in defining an ant based routing algorithm for MWNs. We argue that a routing algorithm based on smart ants designed for mesh networks should have the following desirable properties:

- The smart ants while creating paths should take into account the two types of interferences that inherently exist in mesh networks namely inter-flow interference among the nodes and intra-flow interference along the path of a flow.

- The smart ants should be able to evaluate the load on nodes in order to properly qualify the outgoing links. This would help in *detouring* the packets to new route and hence would result in a more load balanced network.
- Since network nodes in a WMN can be equipped with multiple radios, a smart ant should be able to discover *channel-diverse paths* in order to reduce the interference and effectively improve the overall network throughput.

Smart ants are designed to exhibit the above mentioned desirable properties and it is because of these properties which make our ants smart (intelligent) and that is why we call them *smart ants*. In the following, we show how to incorporate these desirable properties in our smart ant-based routing algorithm (AntMesh).

4. THE ANTMESH ALGORITHM

In this section, we describe the details of AntMesh; a distributed routing algorithm which incorporates smart ants to find high throughput paths with less interference and improved load balancing specifically designed for WMNs. The basic operation of AntMesh follows the routing protocol described in [15, 29]. Smart ants in the form of control packets are generated at regular intervals from each node towards destinations in the network. Indeed, three types of ants are generated: forward smart ants (FSA) which travel from source to destination to discover paths, backward smart ants (BSA) travelling from the destination to the source to update the routing tables and hello smart ants (HSA) which collect the local link quality information to populate link estimation table. In AntMesh, both the FSA and BSA use high priority queues so that the FSA do not need to carry their per hop experienced trip times, rather BSA will estimate their trip time.

4.1 Data Structures

In AntMesh, every node maintains three types of data structures explained as below.

Pheromone table (Probabilistic routing table). This data structure stores the fitness of choosing a specific neighbour as next hop to reach a particular destination in the form of a probability. In other words, it contains pheromone trail information for routing from current node to destination node via next hop. Thus, the pheromone table at a particular node k contains m_k rows where $m_k = |N_k|$ (N_k is the set of neighbouring nodes to node k) and each row contains N columns, where N is the total number of possible destinations in the network (total number of nodes, or population). So an entry P_{id} is the probability of sending a packet to destination d via link i and thus following relation holds for every column in the pheromone table at node k:

$$\sum_{i \in N_k} P_{id} = 1 \qquad \forall d \in [1 \dots N] \qquad (1)$$

Delay table. The second data structure that is maintained by each node is the delay table that stores the average trip time to each destination in the network from the current node (thus this table will have m entries one for each possible destinations in the network). The value stored is an average calculated from the delay value carried by the last W number of smart ants received.

Link estimation table. The third table maintained by AntMesh is the local estimation table which contains the quality/strength of the outgoing links of that particular node to its neighbours. AntMesh uses hello smart ants (HSA) to measure these local link statistics in terms of link level packet transmission delays.

A more detailed explanation on how these data structures are populated will be provided later in this section.

4.2 Node Transition Rule

In order to increase the chances of selecting paths with less interference and more throughput, AntMesh has adopted a pseudo-random node transition rule. A forward smart ant v at an intermediate node k chooses the next hop u to reach to a particular destination d according to Eq. (2) and Eq. (3).

$$u = \begin{cases} \arg\max_{u \in N_k}\{\tau_v(d,u)\} & if\ p \le p_0 \\ P_v(d,u) & otherwise \end{cases} \tag{2}$$

$$P_v(d,u) = \frac{\tau_v(d,u)}{\sum_{i \in N_k} \tau_v(d,i)} \tag{3}$$

where p is a random number uniformly distributed between [0, 1] and p_0 is a constant in the range [0, 1]. Similarly $\tau_v[d,u]$ is the pheromone intensity on the link connecting next hop u with k to reach the destination d. Eq. (2) indicates that if $p \le p_0$, the node with the maximum pheromone value among the neighbours will be selected, otherwise, a proportional selection will be made with the probability $P_v(d,u)$ based on the probability distribution of the neighbours as shown in Eq. (3). Note that the parameter p_0 determines the relative importance of exploitation versus exploration. In AntMesh, since pheromone table is the only data structure that contains the routing information, all the data packets are forwarded based on this pseudo-random node transition rule. Therefore, the high value of p_0 would direct all the traffic to the best path as the link with maximum pheromone value would be selected most of the time and a lower value of p_0 will allow the data traffic to be spread across multiple links thereby resulting in automatic load balancing. How much should this p_0 be set depends highly on the offered load on the mesh network and we will study the effect of this parameter to AntMesh performance in detail in our simulation section. Setting p_0 values to 1 will always forward the packets on links with the best quality.

4.3 Interference Estimation Rule

This rule is designed to fully capture the characteristics of WMNs including the types of interferences that exist in such networks and to incorporate the desired properties of smart ants in AntMesh algorithm. It consists of two modules i.e. link estimation module (LEM) and path estimation module (PEM). These two modules help the smart ants to accurately measure the inter and intra-flow interference in the network.

4.3.1 Link Estimation Module (LEM)

The link estimation module of AntMesh calculates the quality/cost of a wireless link in terms of average transmission time (delay) it takes the MAC layer to send a packet on a particular outgoing link. We define this *link transmission delay* as the time from when a packet starts to be serviced by the MAC layer to the instant that it is successfully transmitted (thus including the time required by any or all retransmissions). Let T_i denote the transmission delay over link i and N_{tx} denote the number of transmissions including retransmissions needed to successfully receive a packet. (In most practical cases, $N_{tx} \cong 1$, as almost all the current wireless devices are equipped with multi-rate feature which automatically adjusts the link rate according to the link quality, resulting in successful transmission of around 90% packets at the first time - see [18]). L_{pkt} is the data packet size and R_s is the link speed. The link transmission delay can be defined as follows:

$$E[T_i] = N_{tx} \times \left(MAC_{oh} + \frac{L_{pkt}}{R_s}\right) \tag{4}$$

where MAC_{oh} is the standard packet sequence of sending a data packet; in IEEE 802.11, MAC_{oh} is calculated as follows:

$$MAC_{oh} = T_{rts} + T_{cts} + 3T_{sifs} + T_{difs} + T_{ack} \tag{5}$$

and T_{rts}, T_{cts}, T_{ack} are the times required for the transmissions of RTS, CTS, and ACK frames respectively, T_{sifs} and T_{difs} are the inter-frame spaces: SIFS and DIFS.

However, this delay $E[T_i]$ in Eq. (4) is just the MAC layer transmission delay which does not take the traffic load into account, i.e., the queuing delay, which depends on the number of packets waiting in the buffer for transmission. The packets that are already in the queue must be served by the MAC layer before the new packet that has just arrived at the node would be served. Therefore, the link estimation module (LEM) calculates the total transmission delay of a packet on a particular outgoing link as the packet transmission delay of the number of packets in the buffer plus the transmission delay of the newly arrived packet. This can be shown as below:

$$LQ_i = E[T_i] \times Q_k + E[T_i] \tag{6}$$

Where Q_k denotes the queue size of node k and thus LQ_i is the link quality of link i. Note that LQ_i includes the load on node k and is therefore the total time it would take for a newly arrived packet at a node k to be transmitted to the next hop.

4.3.2 Path Estimation Module (PEM)

The path estimation module of AntMesh has been designed to meet the remaining desirable properties of smart ants i.e. discovering channel diverse paths and inter/intra-flow network interference.

After the completion of path discovery of forward smart ant (FSA), i.e. it has reached the destination, a backward smart ant (BSA) is created and is sent back to the source to update the data structures maintained by each node. During the backward smart ant's (BSA) travel to the source, each node that encounters the BSA updates its delay table data structure. They do so by estimating the smart ant's trip time from the current node to the destination node. This trip time is estimated as the sum of the average transmission time it takes to send a data packet from the current node to the next hop node (from where the smart ant has arrived) and the estimated accumulated trip time calculated so far by the BSA. The local link transmission delay of the next hop node is provided by the link estimation module from the link estimation table.

Inter-flow Interference. Since, nodes transmitting on the same wireless channel compete for the shared medium, whenever a node is involved in a transmission; its neighbouring nodes should not communicate at the same time with other nodes on the same channel. The PEM of AntMesh captures this type of interference (inter-flow interference) on each node and incorporates it into the local link estimation table. Let $I_{(k)}$ denote the set of queue sizes of node $k's$ interfering nodes. We define $IFLD_{k,i,c}$, "inter-flow link delay" as follows:

$$IFLD_{k,i,c} = LQ_i \times max[I_{(k)}] \tag{7}$$

where $IFLD_{k,i,c}$ is the link transmission delay of a particular link i transmitting on channel c when inter-flow interference is considered. The path estimation module (PEM) only takes the maximum queue size of the neighbouring node among all the neighbours (because a node with a very short queue length can still be congested if its interfering nodes have a lot of packets to send out) as it really depends on the current activities of all neighbouring nodes to find out if they are indeed contending or not. AntMesh uses hello smart ants (HSA) to capture these node statistics in terms of queue sizes among the neighbouring nodes. Therefore a high-contention link that would result in increased link transmission delay is detected timely by the PEM module of AntMesh.

The rationale behind using the queue lengths of the neighbouring nodes to measure the contention on a link belonging to a particular node can be understood with the help of a simple example as illustrated in Figure 1: Let us assume that each node is configured to use a single channel. Let us say that S is the source node and D is the destination node. So there are two paths to reach to D: S-A-D and S-C-D. In this particular diagram, node S calculates the inter-flow interference of its neighbouring nodes A and C by taking into account the queue sizes of the neighbouring nodes of A and C on links i and j respectively. It is clear that neighbouring node A has more neighbours than node C, which initially gives the impression that the contention on link i is going to be high than link j. However, it really depends upon the number of packets waiting to be served in the queues of those neighbouring nodes which describes the contention along the link. So the neighbours around link j which are less as compared to link i

have more packets waiting in the queue than the neighbours of link i. Therefore, the contention on link i is less as compared to link j which is captured by our estimation module and this would eventually result in improving the overall network performance of a mesh network.

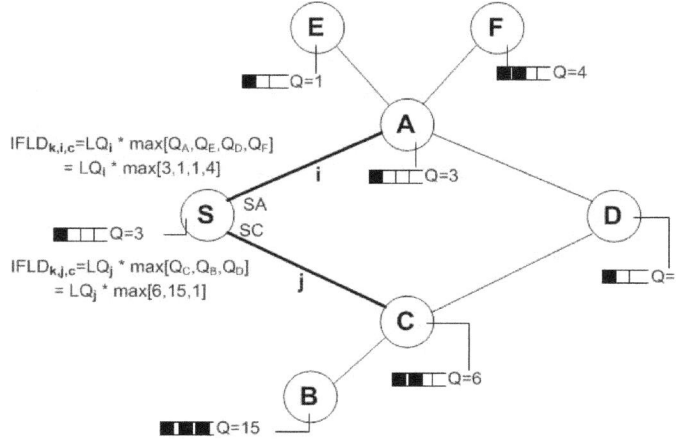

Fig. 1. AntMesh inter-flow interference calculation at node S

Intra-flow Interference. The backward smart ant (BSA) in our smart ant-based routing follows a deterministic path from destination to source; it keeps a record of the channels of the links on which it has traversed together with the accumulated trip time. When the BSA is received by a node (intermediate, or destination) the PEM checks the smart ant for the links it has traversed; if the two latest hops used the same radio channel then it increases the total transmission time calculated in Eq. (7) by a cost factor α as shown below:

$$ITT_{i,c} = IFLD_{k,i,c} \times \alpha \tag{8}$$

where $ITT_{i,c}$ (inter + intra-flow transmission time) is the total link transmission delay of a link i on channel c of a particular node. The cost factor $\alpha\ $is calculated as:

$$\alpha = \frac{2Q_{next}L}{B_i} \tag{9}$$

where Q_{next} is the number of packets waiting in the queue at the next hop node, L is the packet size and B_i is the bandwidth of the link. Since, only one link can be active at any time if both the consecutive links are on the same channel then the effective bandwidth will be $B_i / 2$ (i.e., the transmission time should be doubled for packets using the same channel on consecutive links). Similarly, no cost will be added to the total transmission delay when the two last-hop channels are different along the path ($\alpha = 0$). $ITT_{i,c}$ will be added to the trip time that the BSA has accumulated so far from the next hop neighbour node j to the destination node d denoted as $Trip_{i,d}$. Thus the BSA updates the trip time field of the delay table and hence the total trip time from a node i to destination node d is cumulative and is calculated as:

$$Trip_{i,d} = ITT_{i,c} + Trip_{j,d} \tag{10}$$

where $Trip_{j,d}$ is the trip time from next hop node j to the destination d and is taken from the *Trip Time* field of the BSA packet structure.

4.4 Pheromone Updating Rule

After the calculation of trip times $Trip_{i,d}$, BSAs trigger an update of the pheromone table values corresponding to a specific destination. However, there is a need for a mechanism to correctly integrate the transmission time measured by the AntMesh, into the pheromone table entries. This will improve the routing decision process with respect to the quality of the link. Let us denote the average trip time for a particular destination d stored in the delay table at node i with $T_{i,d}$, and the current calculated trip by $Trip_{i,d}$ then:

$$\Delta p = \frac{1}{2} \times \left(\frac{T_{i,d}}{Trip_{i,d}} \right) \tag{11}$$

where Δp is the reinforcement value that will be added to the pheromone table and it depends upon how good $Trip_{i,d}$ is. Since, AntMesh keeps up to date local link quality information, the trip time $Trip_{i,d}$ depicts the latest network condition in terms of traffic load and congestion.

Now, the pheromone values in the pheromone table corresponding to a particular destination d via next hop i can be updated as follows:

$$P_{i,d} = \frac{P_{i,d} + \Delta p}{1 + \Delta p} \tag{12}$$

Similarly, the other neighbors as next hops j for the same destination can be downgraded:

$$P_{j,d} = \frac{P_{j,d}}{1 + \Delta p} \quad where \; j \neq i \tag{13}$$

Note that Eq. (12) and Eq. (13) satisfy Eq. (1).

4.5 An Illustrative Example

We use a simple wireless network in Figure 2 to illustrate how our proposed smart ant-based routing algorithm works. We denote the source and destination nodes by S and D respectively in this example. CH represents the channel number on which a particular link is configured. The number of packets waiting in the queues are represented next to each node circles. Similarly, the link estimation and delay tables of each node are shown in the figure and the path computation equation is shown above each table to calculate the trip time. Our goal is to find a path from S to D; the path discovery process works as follows: in order to compute our path metric when the forward smart ant (FSA) traverses the network, we need the link metric for each link traversed and the channel of last two hop links in which they are operating. So we overload the smart ant packet to carry the link metric and the channel of links traversed. Following are the sequence of steps performed by the AntMesh algorithm.

1. Node S initiates the route discovery by generating an FSA, which reaches the destination D using the pheromone tables on each node by applying pseudo-random node transition rule.
2. The FSA can reach to destination node D through two paths i.e. S-A-F-D (path 1) or S-G-C-F-D (path 2). Now, destination node D generates the backward smart ant (BSA) which follows the same path as FSA did and would update the pheromone table, local link estimation table and delay table using our custom designed estimation modules of interference estimation rule.
3. When the BSA reaches node F, it records both inter- and intra-flow interferences of the link it has traversed by taking the maximum of the queue sizes of the neighbouring nodes of D (max$[Q_D]$=1) and adding α ($\alpha = 0$) if the last two hops link channels are same respectively. It then incorporates these measurements into the link quality of the link (link estimation module - maintained by link quality table). This step will be the same for either path-1 or -2.
4. For path 1 (S-A-F-D), as the backward smart ant travels to A, it carries with it the accumulated trip time of F-D which it calculated on the previous step. On reaching node A, the inter-flow interference is measured (max$[Q_F,Q_C,Q_D,Q_E]$=5) and then is incorporated into the local link quality of the link A-F stored at link quality table using Eq. (7). Also, since the last two hops' link channels are different (AF = 2, FD = 4), therefore no intra-flow interference exists along these links ($\alpha = 0$). The same calculations are carried for the BSA traversing on path 2 (S-G-C-F-D) i.e. on reaching

node C, the trip time for a packet to reach node D from C is calculated using (12). This trip time is then integrated into the pheromone table using (14, and 15).

5. Step 4 is repeated for node G on path 2 and eventually upon reaching source node S, it captures the inter- and intra-flow interferences of its outgoing link on which the smart ant has arrived (updating the data structures accordingly). Note that for path 1, there exist intra-flow interference along the links SA and AF as they both are on the same channel and this has been taken care by the interference estimation rule using its path estimation module by adding a factor α (fixed to 0.5 in the figure) calculated using Eq. (8).

The example points out that although path-1 contains less number of hops than path-2, the contention in terms of inter- and intra-flow interference along the path is higher. Therefore, the trip time to reach destination D on node S via the outgoing link G is less than through neighbour node A. This is depicted in the delay table of node S in figure 2. This concludes our example on how our smart ants can capture interferences along the routes; we will show that this behaviour will eventually result in an overall improved performance.

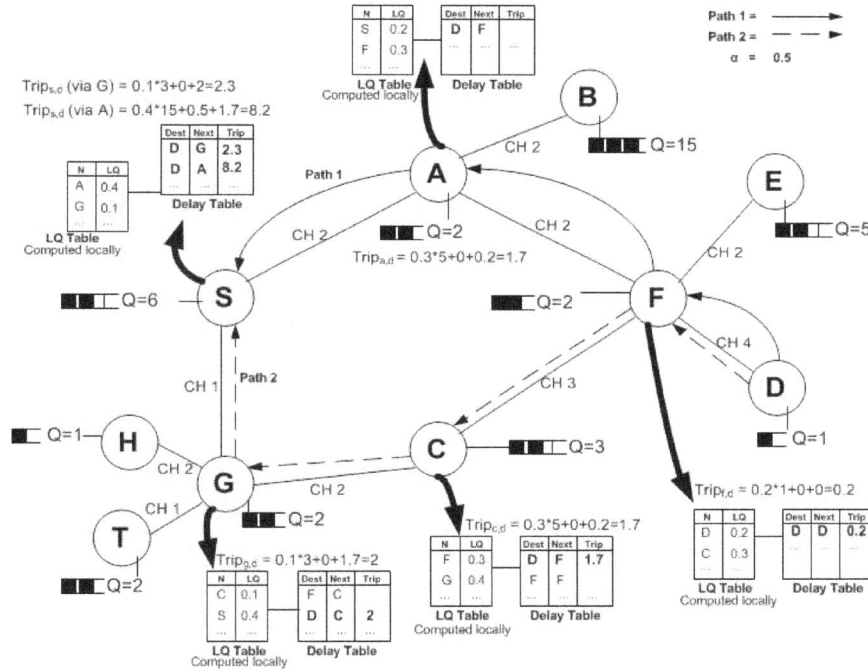

Fig. 2. Illustrative Topology - An example illustrating the working of Smart Ants in AntMesh. The backward smart ant computes trip time and updates delay and pheromone table accordingly. The trip time on each node is calculated using equation: $Trip_{i,d} = LQ_i \times \max[I_k] + \alpha + Trip_{j,d}$ where $\alpha=0.5$ and $Trip_{j,d}$ is the trip time from next hop node to destination, taken from BSA packet

5. PERFORMANCE EVALUATION

In this section, we discuss the implementation, setup, motivation behind the setups, and results of our experiments. Our evaluations are divided into three parts. First, we study the stability of our smart ant-based routing algorithm in terms of how quickly it manages to find new paths in dynamic networks by tuning the algorithm parameters in order to reach an optimized behavior. We also discuss the effect of smart ant generation rate on the overall performance of the mesh network by measuring normalized routing load (NRL) and packet delay. The second part of our evaluation consists of comparing AntMesh with some of the existing well-known routing protocols in WMNs using different network topologies and traffic characteristics. Finally, the performance of AntMesh in mobile WMNs where some of the nodes are static and some of them are mobile is compared in the third part of our evaluation.

5.1 Implementation

In this subsection we will elaborate the implementation details of our proposed smart ant-based routing algorithm (AntMesh) in ns-2 [19]. We have selected ns-2 for implementation and simulation purposes based on its popularity and acceptance in the academic community. We have implemented our AntMesh routing agent as smart ants which are small fixed size control packets being sent periodically for finding paths and building routing and pheromone tables along the way in the network. In order to measure the local link qualities by our link estimation module (LEM) of AntMesh, a list of one-hop neighboring nodes and their corresponding link qualities (i.e., in the form of link delays) are maintained in a local link estimation table by periodically broadcasting hello smart ants (HSA). Upon receiving the HSA, each node sends back its queue size along with the list of its one-hop neighbors and their corresponding queue sizes to calculate the inter-flow interference by the path estimation module (PEM) of link interference rule. This effectively enables each node to capture 2-hop neighborhood information. Periodical HSA type hello exchange has been widely adopted by most of the existing routing protocols [2, 3, 16, 31].

Since current ns-2 distribution does not support multi-channel communication and wireless nodes having multiple interfaces, we included multiple interface support in ns-2 using [36] document by ramon and campo in order to provide support for multi-radio mesh networks and to evaluate the algorithms on such scenarios. Similarly, we have implemented some of the most recent ant-based routing algorithms i.e. SARA [34] and DAR [35] in ns-2 for comparative study. Our selection of these two protocols for comparative evaluation is because these are the most recent routing algorithms proposed that are based on ants to build routing paths. Although, both SARA and DAR are designed for MANETs but the fact that AntMesh is the first ant-based routing approach to deal with WMNs to the best of our knowledge, force us to choose these two protocols. However, this does not imply that MANETs and WMNs have similar characteristics and therefore, a need to design a new routing algorithm for WMNs specially tailored for its unique characteristics was required as mentioned in section I.

Since, most of the traffic in a real WMN is either to or from a wired network [1, 20] (i.e., through Internet gateway points), in our simulations, flows are destined to one to four gateway nodes. The common configuration parameters for all simulation studies in this section are listed in Table II. Although some may argue that some of these values are low (e.g., link bandwidth), our purpose is to provide a comparative evaluation (relative results) and thus we believe that these values will not strongly influence our results. Each of our depicted data points in the results is an average over enough simulation runs to claim a 95% confidence that the relative error of them is less than 5%. Any topology related change in the simulation will be mentioned in the appropriate subsection.

TABLE II
SIMULATION CONFIGURATIONS

Simulation area	$1000 \times 1000 \text{ m}^2$
Transmission range	250m
Propagation model	Two-ray ground
MAC protocol	802.11 CSMA (RTS/CTS disabled)
Link bandwidth	2 Mbps
Traffic type	CBR (UDP)
Packet size	512 bytes
Number of nodes	15 or 100
Number of radios	≤ 3
Hello interval	1 second
Buffer size	20 packets

5.2 Tuning AntMesh Parameters

We study two configuration parameters of AntMesh routing algorithm. The first is the parameter p_0 which is used in our pseudo-random node transition rule and it governs the relative behavior of smart ant forwarding. The parameter p_0 determines the probability of choosing a next hop for path discovery/data forwarding with maximum pheromone value or selecting the next hop based upon probability distributions using Eq. (3). If we set the value of p_0 to be very high, it would select links with high pheromone values more often than others. Similarly, setting the value too low would spread the data traffic around the links and therefore, would fall back to the classical AntNet algorithm. So there exists a relationship and in order to prove this analytical reasoning, figure 4a demonstrates the performance of AntMesh in terms of average network throughput as a function of p_0 under varying network traffic. Notice that the network throughput is almost the same for all the value of p_0 when there is little traffic on the network. The p_0 value starts affecting the algorithm performance when more packets are beginning to pump in the network increasing the network traffic. At that point, the higher the p_0 value, the better the throughput it gives, this is because, most of the time, links of good quality (highest pheromone) are selected for data forwarding. Since the network is still not saturated, the interference estimation rule perfectly captures the interference thereby resulting in increasing the pheromone values of links with less interference. However, notice that when the network traffic is high, the gap between the AntMesh throughput with different values of p_0 became small. We believe the reason behind this pattern is due to the fact that since the network is saturated, the low p_0 value (i.e. 0.2, 0.5) starts spreading the data over multiple links with more chances of random next hop selection. However, for higher values of p_0, AntMesh still gives better throughput than when a lower value is set and therefore, in the rest of our simulations, we will select the value of parameter to p_0=0.8.

The second parameter we intend to tune is the smart ant generation rate which is related to AntMesh stability. The stability of any routing algorithm depends upon how quickly it adapts itself to the changing dynamics of the network. We define the time it takes for routing algorithm to learn the best routing policies as the learning time (convergence time). We demonstrate the stability of our algorithm by measuring *path latency* in mesh networks under dynamic load situations when traffic characteristics change or load on the network is increased. In order to effectively demonstrate the quality of the learned policy, we have used the average packet delay as the evaluation metric.

Figure 4b shows the learning times of AntMesh algorithm under different smart ant generation rates. It relates the protocol's stability in the adaptation process for different ant rates during the simulation run time. We have used a 15-node grid mesh topology shown in figure 3 as the underlying network. In the beginning, one flow is generated to introduce a light load on the network. Then at $t = 10s$, 3 more flows are initiated in order to increase the traffic in the network. At $t = 20s$, these 3 flows are stopped in order to bring the network back to its normal state. It can be seen in figure 4b that AntMesh adapts to this increase in network load at $t=10s$, by switching to a new path between the source and destination node with the help of its link and path estimation modules defined in interference estimation rule. However, at $t = 20s$, the algorithm starts converging back to the previous best path that it had discovered. We believe that the convergence of AntMesh to the best available path is related to how fast smart ants are being generated by each node i.e. the ant generation rate. Figure 4b shows that a more frequent ant generation leads to less the time for the algorithm to find a better path. In other words, the time window for finding this path decreases with increasing number of smart ants in the network. However, these values of various ant rate is highly dependent upon the traffic characteristics as well as the network topology and therefore should not be taken as general criteria for applying AntMesh in WMNs.

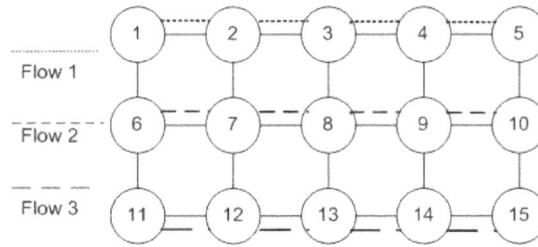

Fig 3. Grid Topology

Although the learning time of AntMesh is inversely proportional to the smart ant generation rate, a higher ant rate means increase control traffic in the network which in turn would limit the overall useful network capacity. Therefore there exists a tradeoff between the amount of routing traffic generated and the learning time. We investigate this tradeoff in AntMesh by using normalized routing load (NRL) as the evaluation metric which is the ratio of the control traffic in the network to the total amount of packets that were actually pumped into the network. Figure 4c shows the NRL when the network is lightly load, and when it is overload under varying smart ants' rates. One interesting thing to note in here is that the network load has almost negligible effect on the routing load when there are too few number of smart ants in the network i.e. the ant generation rate is small. Even for higher ant rates, one can observe that ant generation rates have a marginal effect on the NRL of our routing protocol under the conditions when the network is below saturated. On the other hand, there is a significant difference among the NRL values on the network when the network is overloaded. One reason behind this increase in NRL is due to the increase in packet collisions and the dropping of packets from the node queues when the network becomes overloaded, which in turn will generate more smart ants to discover new paths. Furthermore, there is an extra overhead in communicating queue sizes (2-hop neighbor information) with the neighboring nodes by HSA on each node to compute the inter-flow interference. Although AntMesh has a higher NRL as compared to the overhead of other routing protocols (i.e., distance vector or link state as shown in [12]), we argue that this can be compensated by the efficient performance it provides in routing the traffic by finding less interference paths in WMNs as shown in our results section.

Figure 4d shows the impact on the network performance in terms of end-to-end delay which is depicted as a function of ant generation rates for different network loads. Notice, that the packet delay remains almost the same for increasing number of smart ants in the network.. This is because most of the traffic stays on the best path which keeps the packet delay almost constant among different network load conditions. It is because of this reason, for the rest of our simulations, we select the ant generation rate to 40 ants per second.

5.3 Results from WMNs - Stationary Nodes

We want to observe the effectiveness of our proposed smart ant-based routing algorithm in capturing inter and intra-flow interference on a mesh network having multiple radios each configured to multiple channels. For evaluating AntMesh in a multi-radio WMN environment, we modified the node models in our studies to contain one to three radios, where each radio may be configured to work with multiple channels. 802.11b DCF with RTS/CTS disabled is used as the underlying MAC protocol with link bandwidth of 2Mbps. We have run our experiments on an infrastructure grid mesh topology (as in Figure 3) with random location nodes added. In this semi-random topology we place 20 nodes uniform randomly in a 1000m by 1000m area, and four traffic flows are generated destined to 1 to 4 internet gateway nodes.

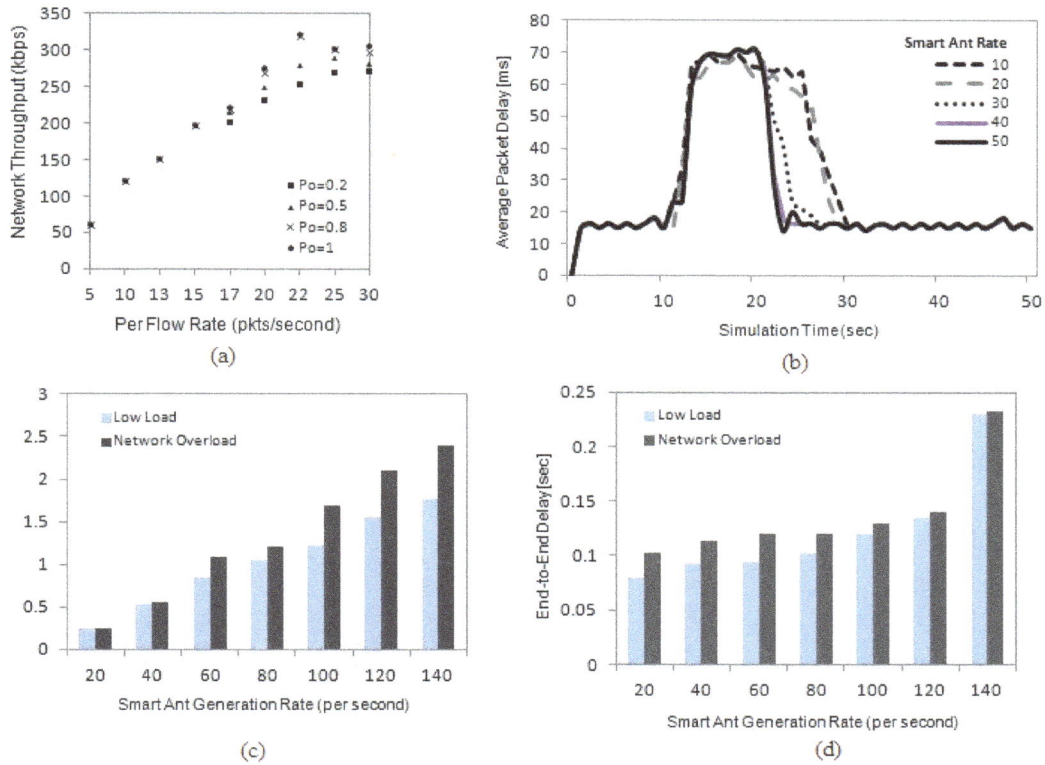

Fig. 4. AntMesh behavior (a) throughput as a function of p_0. (b) learning time with various smart ant rates. (c) Effect of smart ant generation on protocol overhead. (d) Effect of smart ant generation on AntMesh end-to-end delay

Figure 5a presents the total network throughput vs. the flow rate for our proposed smart ant-based routing algorithm. We can observe that AntMesh provides better throughput, outperforming the other schemes by as much as 35% especially when the network becomes congested. Both the SARA and DAR perform poor against AntMesh but almost equally with each other since it does not capture the traffic load among the nodes. The increased throughput of AntMesh is due to the capturing of inter- and intra-flow interference by its interference estimation rule. Although, DAR performs close to AntMesh when the network load is light, as the network starts to saturate, nodes' queues start filling up, experiencing more contention on the outgoing links thus affecting the link quality; this is only recognized by smart ants with the help of their path estimation module (PEM) of interference estimation rule (by taking into consideration the queue sizes being periodically sent by HSA to all the neighbors of node). The involvement of the PEM module of AntMesh results in earlier detouring of the flow than with the other schemes. Clearly, the benefit of considering the queue sizes of neighboring nodes to measure the actual interference and traffic load is evident from the increased throughput in the graph.

Similarly, Figure 5b demonstrates the average end-to-end delay. AntMesh outperforms other schemes in terms of end-to-end delay of packets by as much as 30%. Notice though, that the end-to-end delay of DAR approached very close to AntMesh when the network is heavily loaded; this is because of overhead experienced by AntMesh due to the path change captured by the stochastic behaviour of queue lengths.

The packet loss ratios of all three approaches are shown in Figure 5c. When the network load is low, the packet loss ratios, as expected, for all of the three schemes are low; as the network becomes more and more saturated, AntMesh outperforms SARA and DAR by avoiding inter- and intra-flow interference paths. Particularly, the difference in packet loss ratio among

AntMesh and other two schemes is very evident and can result in as much as a 60% reduction.

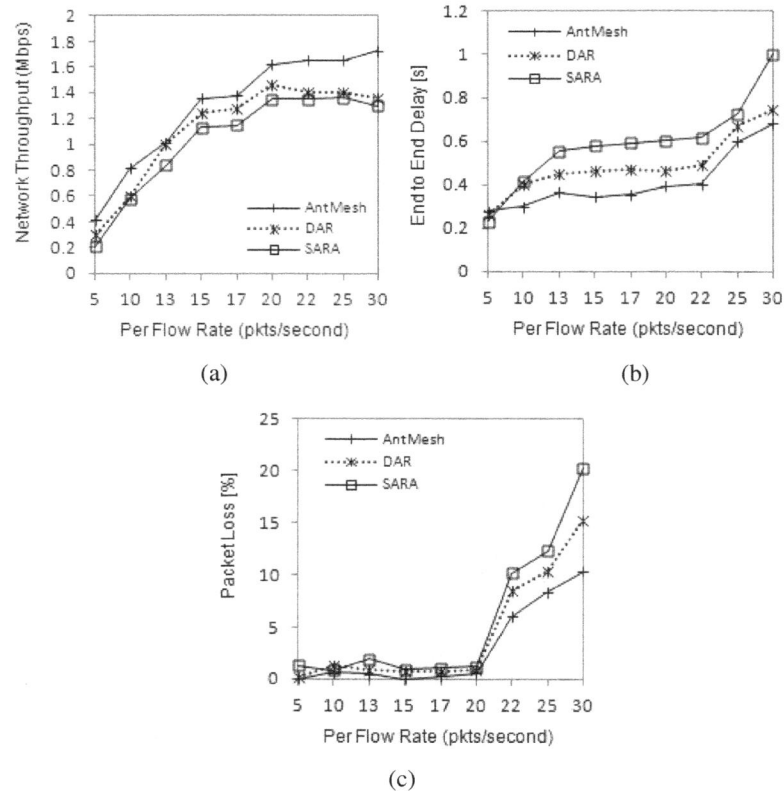

(a)   (b)

(c)

Fig. 5. AntMesh (a) Network throughput. (b) End-to-End delay. (c) Packet Loss

5.4 Results from WMNs - Mobile Nodes

In this part of our evaluation, we test the performance of AntMesh in mobile WMNs scenario with static and mobile nodes. We consider a network of 100 nodes randomly distributed in a 500m x 500m region. All the mobile nodes in each simulation run of 60s are configured with the same speed and the random waypoint model [33]. The data traffic consists of 6 CBR flows of random source destination pair. In order to evaluate the effectiveness of our smart ant-based routing algorithm, we have used packet delivery fraction and average end-to-end delay as evaluation metrics. Figures 6a and 6b show the delivery ratio and average end-to-end delay as a function of different node speeds ranging from 0 m/s to 30 m/s. As it can be easily seen from the delivery ratio graph, AntMesh outperforms SARA and DAR clearly and this performance gap between the schemes is more evident when the nodes are moving at a faster rate. However, this performance gap for average packet delay in Figure 6b is less but again it increases for higher nodes' speeds. Notice the dip in the delivery ratio for all the algorithms in Figure 6a, we believe that this is because since nodes are moving in fast speeds, it is possible that some nodes get out of reach from the rest of the network and therefore packets cannot be delivered to them resulting in possible low delivery ratio. Similar arguments can be made for the sharp rise in packet delay of the algorithms in Figure 6b. Furthermore, in the mobility model (RWP) that we have used in our experiments, nodes tend to make sudden and uncorrelated changes in their movement direction at the pause points, which is captured timely by our adaptive smart ant-based routing algorithm resulting in improved performance than the SARA and DAR protocols which are reactive in nature. Another interesting result is that SARA performs better than DAR and very close to AntMesh in both the graphs. This is due to the design objective of SARA which is targeted for networks having highly mobile nodes with critical connectivity. However, it still under performs than AntMesh because of its inability in effectively measuring the interference.

In order to simulate the true nature of a wireless mesh network with mobile nodes, we have collected another set of results by gradually increasing the number of mobile nodes in the network among the fixed nodes in the networks. The results are collected with 20% to 100% of the total nodes in the network given a random movement and direction while the remaining was stationary. We have provided the speed as a simulation parameter that was fixed to 10 m/s. Figures 6c and 6d show the same performance measures for all algorithms as a function of number of mobile nodes in the network. It can be seen from the graphs that AntMesh exhibits a better performance in terms of higher delivery ratio and less packet delay than SARA and DAR particularly when the network is highly mobile. The stochastic forwarding of smart ants for path exploration and timely capture of inter and intra flow interference by our link and path estimation modules of smart ants to pay off more when the number of mobile nodes in the network are increased. The reason for the comparatively poor performance of SARA and DAR lies in the fact that since it uses minimum hop count metric for path construction, therefore, it fails to cope up with the sudden and dynamic changes in the network until the path gets broken which would eventually result in increasing the packet delay due to frequent path reconstructions as shown in Figure 6d. Also, notice that in these graphs too, the performance of SARA is better than that of DAR because of the previously mentioned design objective of SARA.

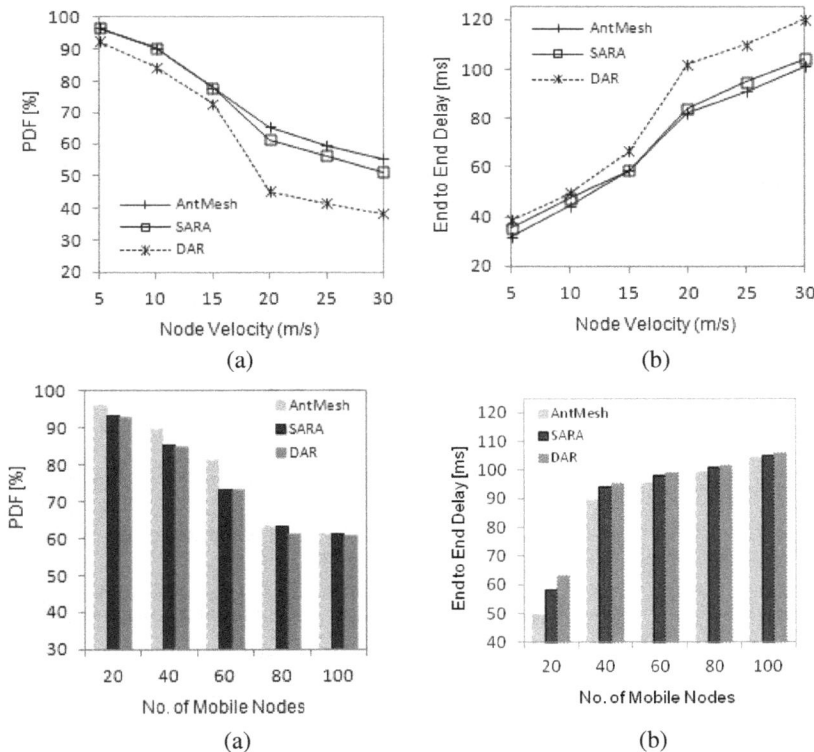

Fig. 6. AntMesh (a) PDF as a function of node speed (b) Packet Delay as a function of node speed (c) PDF as a function of mobile nodes (d) Packet Delay as a function of mobile nodes

6. CONCLUSION

This paper studied the problem of packet routing in wireless mesh networks with a specific emphasis on a framework using *smart ants*. To enable the use of such agents we proposed an interference-aware data forwarding scheme called AntMesh which provides a distributed, stochastic heuristic to solve a dynamic network routing problem. In addition, we also emphasized on the importance of having certain desirable properties that smart ants should possess in order to effectively utilize the space/channel diversity typically common in such networks. We demonstrated the stability of AntMesh through simulations that it quickly converges to the best path under situations when traffic characteristics change (among others when load on the network is increased). We have shown that with an appropriate tuning of the parameters, AntMesh behaves better when compared to other competing approaches in mesh networks. In addition, the promising results shown in the paper underline the need for a real-life testbed evaluation on which we are currently working on.

REFERENCES

[1]   A. Raniwala and T.-C. Chiueh, "Architecture and algorithms for an IEEE 802.11-based multi-channel wireless mesh network," Proc. of the IEEE INFOCOM, 2005, pp. 2223-34.

[2]   D.S.J. Couto, D. Aguayo, J. Bicket, and R. Morris, "A high-throughput path metric for multi-hop wireless routing," Proc. of the ACM MOBICOM, 2003, vol. 1, pp. 134-146, Sep. 2003.

[3]   R. Draves, J. Padhye, and B. Zill, "Routing in Multi-Radio, Multi-Hop Wireless Mesh Networks," Proc. of the ACM MOBICOM, 2004, pp. 114-128.

[4]   Y. Yang, J. Wang, and R. Kravets, "Interference-aware load balancing for multihop wireless networks," Tech. Rep. UIUC DCS-R-2005-2665, Dept. of Computer Science, UIUC, 2005.

[5]   A. P. Subramanian, M. M. Buddhikot, and S. C. Miller, "Interference aware routing in multi-radio wireless mesh networks," IEEE Workshop on Wireless Mesh Networks (WiMesh), Sept. 2006, pp. 55–63.

[6]   M. Dorigo and G. Di Caro, "The ant colony optimization meta-heuristic," in D. Corne, M. Dorigo, and F. Glover, editors, New Ideas in Optimiziation, pages 11-32. MacGraw Hill, 1999.

[7]   M. Dorigo, G. Di Caro, and L.M. Gambardella, "Ant algorithms for discrete optimization," Arfitcial Life, 5(2): 137-172, 1999.

[8]   G. Di Caro and M. Dorigo, "AntNet: distributed stigmergetic control for communications networks," Journal of Artificial Intelligence Research (JAIR), 1998.

[9]   R. Schoonderwoerd, O. Holland, J. Bruten, and L. Rothkrantz, "Ants for load balancing in telecommunication networks," HP Lab., Bristol, U.K., Tech. Rep. HPL-96-35, 1996.

[10] S. Kamali, J. Opatrny, "POSANT: a position based ant colony routing algorithm for mobile ad-hoc networks," Elsevier Wireless and Mobile Communications, pp 21–21, 2007.

[11] M. Gunes, U. Sorges, I. Bouazizi, "ARA-the ant-colony based routing algorithm for MANETs," Proc. International Conference on Parallel Processing, 2002.

[12] G. Di Caro, D. Gambardella, "AntHocNet: an adaptive nature-inspired algorithm for routing in mobile ad hoc networks," European Transactions on Telecommunications, 2005, vol. 16, no. 5, pp. 443–455, 2005.

[13] G. Di Caro and M Dorigo, "Two ant colony algorithms for best-effort routing in datagram networks," Proc. of the 10th IASTED Int., 1998, pp. 541–546.

[14] B. Baran and R. Sosa, "A new approach for AntNet routing," Proceedings of the 9th Int. Conf. Computer Communications Networks, Las Vegas, NV, 2000.

[15] F. Bokhari and G. Záruba, "AMIRA: interference-aware routing using ant colony optimization in wireless mesh networks," Proceedings of the IEEE Wireless Communications and Netwrking Conference (WCNC), April. 2009.

[16] Y. Yang, J. Wang, and R. Kravets, "Designing routing metrics for mesh networks," Proc. of IEEE Workshop on Wireless Mesh Networks (WiMesh), June 2005.

[17] Y. Shi, Y. T. Hou, J. Liu, and S. Kompella, "How to correctly use the protocol interference model for multi-hop wireless networks," Proceedings of MobiHoc, USA, May 2009.

[18] P. Gopalakrishnan, P Spasojevic, L. Greenstein, I.A. Seskar, "Method for predicting the throughput characteristics of rate-adaptive wireless LANs," VTC2004-Fall. 2004, pp. 4528-4532

[19] Fall and Varadhan, "NS notes and documentation," in The VINT Project, UC berkely, LBL, USC/ISI, and Xerox PARC, 1997.

[20] I.F. Akyildiz, X. Wang, and W. Wang, "Wireless mesh networks: a survey", Computer Networks and ISDN Systems, v.47 n.4, p.445-487, 15 March 2005

[21] M. Dorigo & L.M. Gambardella, "Ant Colony System: A Cooperative Learning Approach to the Traveling Salesman Problem," in IEEE Transactions on Evolutionary Computation, 1(1):53-66, 1997

[22] I. A. Wagner, M. Lindenbaum, A. M. Bruckstein, "ANTS: Agents, Networks, Trees, and Subgraphs," Future Generation Comp. Systems journal, North Holland June 2000

[23] Perkins and Royer, "Ad hoc On-demand Distance Vector Routing." In IEEE Workshop on Mobile Computing and Systems and Applications, 1999.

[24] Philippe Jacquet, Paul Muhlethaler, Amir Qayyum, Anis Laouiti, Laurent Viennot and Thomas Heide Clausen "Optimized Link-State Routing Protocol", draft-ieft-olsr-04.txt - work in progress, March 2001.

[25] X. Wang, F. Li, S. Ishihara, T. Mizuno, A multicast routing algorithm based on mobile multicast agents in ad-hoc networks, Special Issue on Internet Technology, IEICE Transactions on Communications E84-B (8) (2001).

[26] Y.J. Suh, H.S. Shin, D.H. Kwon, An efficient multicast routing protocol in wireless mobile networks, Wireless Networks 7 (2001) 443–453.

[27] R.R. Choudhury, S. Bandyopadhyay, K. Paul, A distributed mechanism for topology discovery in ad hoc wireless networks using mobile agents, in: Proceedings First Annual Workshop on Mobile Ad Hoc Networking Computing, MobiHOC, Boston, MA, USA, August 2000.

[28] R.R. Choudhury, K. Paul, S. Bandyopadhyay, MARP: a multi-agent routing protocol for mobile wireless ad hoc networks, Autonomous Agents and Multi-Agent Systems 8 (2004) 47–68

[29] Fawaz Bokhari, Gergely Zaruba, "AntMesh: An Efficient Data Forwarding Scheme for Load Balancing in Multi-Radio Infrastructure Mesh Networks" Proceedings of 4th IEEE International Workshop on Enabling Technologies and Standards for Wireless Mesh Netowrking (MeshTech'10) held in conjunction with IEEE MASS, San Francisco, California, November 8-12, 2010

[30] L.M. Gambardella, M. Dorigo, Solving symmetric and asymmetric tsps by ant colonies, in: Proceedings IEEE International Conference on Evolutionary Computation, Nagoya, Japan, May 1996, pp. 622–627.

[31] Y. Yang and R. Kravets, "Contention-Aware Admission Control for Ad Hoc Networks," IEEE Trans. Mobile Computing, vol. 4, no. 1, pp. 363-377, July/Aug. 2005.

[32] AntNet implementation for wired networks, http://antalgorithm.googlecode.com/files/antnet.tar.gz

[33] Johnson DB, Maltz DA. *Mobile Computing,* Chapter Dynamic Source Routing in Ad Hoc Wireless Networks. Kluwer: Norwell, MA, 1996; 153-181

[34] L. Rosati, M. Berioli, G. Reali, "On ant routing algorithms in ad hoc networks with critical connectivity," Elsevier Ad Hoc networks, pp 827–859, 2008.

[35] F. Correia, T. Vazao, "Simple ant routing algorithm strategies for a (Multipurpose) MANET model," Elsevier Ad Hoc networks, pp 810–823, 2010.

[36] Adding multiple interface support in NS-2, http://personales.unican.es/aguerocr/files/ucMultiIfacesSupport.pdf

3

MIMO Interference Management Using Precoding Design

Martin Crew[1], Osama Gamal Hassan[2] and Mohammed Juned Ahmed[3]

[1]University of Cape Town, South Africa
martincrew@topmail.co.za
[2]Cairo University, Egypt
[3]King Abdullah University of Science and Technology, Saudi Arabia

ABSTRACT

In this paper, we investigate how to design precoders to achieve full diversity and low decoding complexity for MIMO systems. First, we assume that we have 2 transmitters each with multiple antennas and 2 receivers each with multiple antennas. Each transmitter sends codewords to respective receiver at the same time. It is difficult to handle this problem because of interference. Therefore, we propose an orthogonal transmission scheme that combines space-time codes and array processing to achieve low-complexity decoding and full diversity for transmitted signals. Simulation results validate our theoretical analysis.

KEYWORDS

Z Channel, Alamouti Codes, MIMO, Interference Cancellation, Complexity, Co-channel Interference.

I. INTRODUCTION

Multiple-input multiple-output (MIMO) channels arise in many different scenarios such as when a bundle of twisted pairs in digital subscriber lines (DSLs) is treated as a whole, when multiple antennas are used at both sides of a wireless link, or simply when a frequency-selective channel is properly modeled by using, for example, transmit and receive filterbanks. In particular, MIMO channels arising from the use of multiple antennas at both the transmitter and at the receiver have recently attracted significant interest because they provide an important increase in capacity over single-input single-output (SISO) channels under some uncorrelation conditions [1–6].

Recently, several space-time processing techniques have been used in multiple access channels to reduce the decoding complexity and enhance system performance by cancelling the interference from different users [7–10]. When it comes to Z channels [11], a scenario when there are two users each transmitting different codewords to two receivers simultaneously, how to achieve low-complexity decoding and high performance such as full diversity is still an open problem.

In this paper, we investigate how to achieve the low-complexity decoding and the highest possible diversity to improve the transmission quality for space-time codes in Zrate feedback channel. This is not unreasonable; control channels are often available to implement power control, adaptive modulation, and certain closed-loop diversity modes.

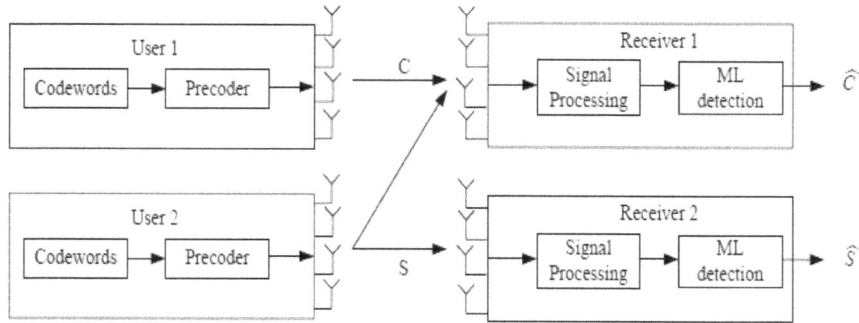

Figure 1: Z Channel

channels without losing symbol rate. We assume that our system operates under short term power constraints, fixed codeword block length and limited delay. Under these constraints, there will always be some outage probability [12–15]. For example, [14] shows that outage probability exisits for the block-fading channel with limited delay and block length. [15] points out that when the delay is finite, for any finite rate, as small as it may be, there is a nonzero outage probability independent of the code length. Thus, the diversity is an important tool to evaluate the system performance.

The outline of the paper follows next. Section II introduces our motivation and the Z channels we discuss in this paper. In Section III, we propose an orthogonal transmission scheme which is necessary to achieve low complexity decoding, high coding gain and full diversity as shown in later sections. In Section IV, our decoding scheme is proposed. We analyze the performance of our scheme in Section V. Simulation results are presented in Section VI and Section VII concludes the paper.

Notation: We use boldface letters to denote matrices and vectors, super-scripts $(\cdot)T$, $(\cdot)*$, $(\cdot)\dagger$ to denote transpose, conjugate and transpose conjugate, respectively. We denote the element in the ith row and the jth column of matrix $\mathbf{X}$ by $X(i, j)$. We denote the jth column of a matrix $\mathbf{X}$ by $\mathbf{X}(j)$.

II Motivation and Channel Model

We assume there are 2 users each with 4 transmit antennas and 2 receivers each with 4 receive antennas. Both users want to send different space-time codes to Receivers 1 and 2 on the same frequency band at the same time. As shown in Figure 1, User 1 wants to send codeword C to Receiver 1 without causing interference to Receiver 2. User 2 wants to send codeword S to Receiver 2 and causes interference to Receiver 1. When channel knowledge is not available at the transmitters, space-time codes combined with TDMA can be used to achieve symbol-by-symbol decoding and full diversity. But the symbol rate reduces to one half. A solution to keep the symbol rate unchanged when space-time codes are used, is to combine space-time coding and array processing. In other words, we allow all transmitters to send space-time codes simultaneously to keep rate one and utilize special array processing techniques to achieve low-complexity decoding and full diversity. In this paper, we achieve the above goals under short-term power constraints, fixed codeword block length and limited delay, when there is outage. We do not claim that our scheme can achieve capacity or full degree of freedom. After all, there is a tradeoff between diversity and multiplexing gain, which is outside scope of this paper. We introduce input-output equations. We let each user transmit Alamouti Codes [17] as follows:

$$\mathbf{C} = \begin{pmatrix} c_1 & -c_2^* \\ c_2 & c_1^* \end{pmatrix}, \quad \mathbf{S} = \begin{pmatrix} s_1 & -s_2^* \\ s_2 & s_1^* \end{pmatrix} \tag{1}$$

where $i, j = 1, 2$. Note that we can also use other space-time codes with rate one and Alamouti code is just one example. Let

$$\mathbf{A}^t = \mathbf{A}^t(4, 4) \tag{2}$$

be the precoders we need to design. In this paper, we use (i, j) denote a matrix of $i \times j$ dimension. They are combined with the space-time codes sent by User 1 and this is the first step of our array processing technique. Note that in order to satisfy the short-term power constraint, we need

$$\|\mathbf{A}^t(4, 4)\|_F^2 = 1 \tag{3}$$

Similarly, the precoders for User 2 is defined as

$$\mathbf{B}^t = \mathbf{B}^t(4, 4) \tag{4}$$

with the power constraint

$$\|\mathbf{B}^t(4, 4)\|_F^2 = 1 \tag{5}$$

The channels are quasi-static flat Rayleigh fading and keep unchanged during two time slots. Then we let

$$\mathbf{H}_l = \mathbf{H}_l(4, 4) \tag{6}$$

denote the channel matrix between User 1 and Receivers l, respectively. Similarly, we use

$$\mathbf{G}_l = \mathbf{G}_l(4, 4) \tag{7}$$

to denote the channel matrix between User 2 and Receiver l, respectively. Then the received signals at Receiver 1 at time slot t can be denoted by

$$\mathbf{y}_1^t(4, 1) = \mathbf{H}_1(4, 4)\mathbf{A}^t(4, 4)\mathbf{C}(t) + \mathbf{G}_1(4, 4)\mathbf{B}^t(4, 4)\mathbf{S}(t) + \mathbf{n}_1^t(4, 1) \tag{8}$$

Where

$$\mathbf{y}_1^t = \mathbf{y}_1^t(4, 1), \quad \mathbf{n}_1^t = \mathbf{n}_1^t(4, 1) \tag{9}$$

denote the received signals and the noise at Receiver 1, respectively, at time slot t. Similarly, at time slot t, Receiver 2 will receive the following signals

$$\mathbf{y}_2^t(4, 1) = \mathbf{G}_2(4, 4)\mathbf{B}^t(4, 4)\mathbf{S}(t) + \mathbf{n}_2^t(4, 1) \tag{10}$$

Where

$$\mathbf{y}_2^t = \mathbf{y}_2^t(4, 1), \quad \mathbf{n}_2^t = \mathbf{n}_2^t(4, 1) \tag{11}$$

Equations (8) and (10) are the channel equations on which we will base our design in this paper.

III Precoder Design and Orthogonal Transmission Structure

In this section, we will build an orthogonal transmission structure by combining the space-time codes and our precoders. This orthogonal transmission structure is necessary because it provides two benefits. The first benefit is that low-complexity decoding can be realized because under this orthogonal transmission structure, different codewords will be sent along different orthogonal vectors. We can easily decode the symbols without the interference at each receiver. The second benefit is that we can achieve full diversity and higher coding gain once we make the proper array processing as shown in later sections. This is the key difference between our array processing method and the interference alignment method. The latter can only achieve the first benefit. Of course, the tradeoff is that we lose the maximum possible degree of freedom in the process.

Different users and different codewords may have different diversities. By saying full diversity, we mean the diversity is full for each codeword sent by each user. For example, full diversity for User 1 means at Receiver 1, the diversity for codeword $\mathbf{C}$ is full. Similarly, by saying the diversity for User 2 is full, we mean that at Receiver 2, the diversity for codeword $\mathbf{S}$ is full. In this section, we show how to build the orthogonal transmission structure by designing proper precoders. Later, we will show that our proposed orthogonal transmission scheme can achieve low-complexity decoding and full diversity.

Our main idea to build the orthogonal transmission structure is to adjust each signal in the signal space of Z channels by using precoders for each transmitter, such that at the receiver each desired signal is orthogonal to all other signals. In Equation (8), we use

$$\mathbf{H}_{11}^{t}(4,4) = \mathbf{H}_{1}(4,4)\mathbf{A}^{t}(4,4), \mathbf{G}_{11}^{t}(4,4) = \mathbf{G}_{1}(4,4)\mathbf{B}^{t}(4,4) \tag{12}$$

to denote the equivalent channel matrices. Then Equation (8) becomes

$$\mathbf{y}_{1}^{t}(4,1) = \mathbf{H}_{11}^{t}(4,4)\mathbf{C}(t) + \mathbf{G}_{11}^{t}(4,4)\mathbf{S}(t) + \mathbf{n}_{1}^{t}(4,1) \tag{13}$$

Similarly, in Equation (10), if we use

$$\mathbf{G}_{21}^{t}(4,4) = \mathbf{G}_{2}(4,4)\mathbf{B}^{t}(4,4) \tag{14}$$

to denote the equivalent channel matrices, we have

$$\mathbf{y}_{2}^{t}(4,1) = \mathbf{G}_{21}^{t}(4,4)\mathbf{S}(t) + \mathbf{n}_{2}^{t}(4,1) \tag{15}$$

By Equation (13), since the receiver has 2 receive antennas, each symbol is actually transmitted along a 2-dimensional vector in a 2-dimensional space. Because each user sends two symbols at the same time, at the receiver, there are 4 signal vectors in the two-dimensional space.

Since we want to send $\mathbf{C}$ and $\mathbf{S}$ along orthogonal directions, we let each one of $\mathbf{C}$, $\mathbf{S}$ occupy only one dimension. In other words, for any codeword, we should transmit each of the corresponding four symbols in the same direction. In this way, there are only 2 transmit directions. Once we can align the 2 transmit directions of $\mathbf{C}$, $\mathbf{S}$ properly, we can separate them completely. This is the main idea to build the orthogonal transmission structure. Note that this is only a general idea and much details are omitted. For example, we will show later that after some array processing and moving the interference at the receiver, each symbol at each receiver will have its own direction. We need to do additional array processing to reduce the decoding complexity and achieve full diversity.

In this section, we only explain the above main idea. By Equation (13), c_1, c_2 are transmitted along $\mathbf{H}_{11}^{t}(4,4)(1), \mathbf{H}_{11}^{t}(4,4)(2)$, respectively. In order to make $\mathbf{H}_{11}^{t}(4,4)(1), \mathbf{H}_{11}^{t}(4,4)(2)$ along the same direction, by Equation (12), we need

$$\mathbf{A}^{t}(4,4)(1) = \frac{1}{\alpha^{t}}\mathbf{A}^{t}(4,4)(2) \tag{16}$$

where α_{11}^{t} is a constant that we will determine later. From $\|\mathbf{A}^{t}(4,4)\|_{F}^{2} = 1$, we know

$$\|\mathbf{A}^{t}(4,4)(1)\|_{F}^{2} = \frac{1}{1+(\alpha^{t})^{2}} \tag{17}$$

So when we design precoder $\mathbf{A}^t(4,4)$, Equations (16) and (17) should be satisfied. Similarly, precoders $\mathbf{B}^t(4,4)$ should also satisfy the following conditions:

$$\mathbf{B}^t(4,4)(1) = \frac{1}{\beta^t}\mathbf{B}_1^t(4,4)(2) \tag{18}$$

with

$$\|\mathbf{B}^t(4,4)(1)\|_F^2 = \frac{1}{1+(\beta^t)^2} \tag{19}$$

Now Equations (13) and (15) become

$$\mathbf{y}_1^t(4,1) = [\mathbf{H}_{11}^t(4,4)(1), \mathbf{H}_{11}^t(4,4)(1)] \cdot \mathbf{C}(t) +$$
$$[\mathbf{G}_{11}^t(4,4)(1), \mathbf{G}_{11}^t(4,4)(1)] \cdot \mathbf{S}(t) + \mathbf{n}_1^t(4,1) \tag{20}$$

and

$$\mathbf{y}_2^t(4,1) = [\mathbf{G}_{21}^t(4,4)(1), \mathbf{G}_{21}^t(4,4)(1)] \cdot \mathbf{S}(t) + \mathbf{n}_2^t(4,1) \tag{21}$$

where $\mathbf{H}_{11}^t(4,4)(1)$, $\mathbf{G}_{11}^t(4,4)(1)$, $\mathbf{G}_{21}^t(4,4)(1)$ denote the first column of matrix $\mathbf{H}_{11}^t(4,4)$, $\mathbf{G}_{11}^t(4,4)$, $\mathbf{G}_{21}^t(4,4)$, respectively. At receiver one, after we combine the channel equations in two time slots, we have

$$\mathbf{y}_1(4,1) = \begin{pmatrix} \mathbf{H}_{11}^1(4,4)(1) & \mathbf{H}_{11}^1(4,4)(1) \\ (\mathbf{H}_{11}^2(4,4)(1))^* & -(\mathbf{H}_{11}^2(4,4)(1))^* \end{pmatrix} \cdot \begin{pmatrix} c_1 \\ c_2 \end{pmatrix} +$$
$$\begin{pmatrix} \mathbf{G}_{11}^1(4,4)(1) & \mathbf{G}_{11}^1(4,4)(1) \\ (\mathbf{G}_{11}^2(4,4)(1))^* & -(\mathbf{G}_{11}^2(4,4)(1))^* \end{pmatrix} \cdot \begin{pmatrix} s_1 \\ s_2 \end{pmatrix} + \mathbf{n}_1(4,1) \tag{22}$$

where

$$\mathbf{y}_1(4,1) = \begin{pmatrix} \mathbf{y}_1^1(4,1) \\ (\mathbf{y}_1^2(4,1))^* \end{pmatrix} \tag{23}$$

and

$$\mathbf{n}_1(4,1) = \begin{pmatrix} \mathbf{n}_1^1(4,1) \\ (\mathbf{n}_1^2(4,1))^* \end{pmatrix} \tag{24}$$

Similarly, at receiver two, after we combine the channel equations in two time slots, we have

$$\mathbf{y}_2(4,1) = \begin{pmatrix} \mathbf{G}_{21}^1(4,4)(1) & \mathbf{G}_{21}^1(4,4)(1) \\ (\mathbf{G}_{21}^2(4,4)(1))^* & -(\mathbf{G}_{21}^2(4,4)(1))^* \end{pmatrix} \cdot \begin{pmatrix} s_1 \\ s_2 \end{pmatrix} + \mathbf{n}_2(4,1) \tag{25}$$

where

$$\mathbf{y}_2(4,1) = \begin{pmatrix} \mathbf{y}_2^1(4,1) \\ (\mathbf{y}_2^2(4,1))^* \end{pmatrix} \tag{26}$$

and

$$\mathbf{n}_2(4,1) = \begin{pmatrix} \mathbf{n}_2^1(4,1) \\ (\mathbf{n}_2^2(4,1))^* \end{pmatrix} \tag{27}$$

By Equation (22), we can see that once we make vector $\mathbf{H}_{11}^1(4,4)(1)$ orthogonal to $\mathbf{G}_{11}^1(4,4)(1)$ at time slot 1 and $\mathbf{H}_{11}^2(4,4)(1)$ orthogonal to $\mathbf{G}_{11}^2(4,4)(1)$ at time slot 2, signal vectors for c_1, c_2 will lie in a subspace which is orthogonal to the subspace created by the signal vectors for s_1, s_2. Because of this orthogonality, at the receiver one, we can easily separate the desired signals c_1, c_2 from the interference signals s_1, s_2. At Receiver 2, since there is no interference, by Equation (25), we can easily decode the desired signals s_1, s_2. This is our main idea to achieve interference-free transmission in this interference channel.

Now we show how to derive the above orthogonality by designing precoders for Users 1 and 2 simultaneously. Assume the Singular Value Decomposition of channel matrices $\mathbf{H}_1(4,4)$, $\mathbf{G}_2(4,4)$ as follows

$$\mathbf{H}_1(4,4) = \mathbf{V}_{H_1(4,4)}\boldsymbol{\Lambda}_{H_1(4,4)}\mathbf{U}^{\dagger}_{H_1(4,4)} \tag{28}$$

$$\mathbf{G}_2(4,4) = \mathbf{V}_{G_2(4,4)}\boldsymbol{\Lambda}_{G_2(4,4)}\mathbf{U}^{\dagger}_{G_2(4,4)} \tag{29}$$

At time slot 1, we let User 1 transmit along its best direction. In this case, we can choose the precoder

$$\mathbf{A}^1(4,4) = \frac{1}{\sqrt{2}}[\mathbf{U}_{H_1(4,4)}(4,4)(1), \mathbf{U}_{H_1(4,4)}(4,4)(1)] \tag{30}$$

Then we design precoders for User 2 such that at both receivers 1, the signal vectors from User 2 are orthogonal to the signal vectors of User 1. Note that at receiver one, the signal from User 2 is interference and at receiver two, the signal from User 2 is the desired signal. We need to consider the signals at both receivers when we design precoder for User 2. So, at time slot 1, the precoder $\mathbf{B}^1$ needs to satisfy the following three equations. At receiver one, we need

$$(\mathbf{H}^1_{11}(4,4)(1))^{\dagger} \cdot \mathbf{G}^1_{11}(4,4)(1) = 0 \tag{31}$$

and the power constraint

$$||\mathbf{B}^1(4,4)(1)||^2 = \frac{1}{2} \tag{32}$$

By solving the above two equations, we can get the precoder $\mathbf{B}^1(4,4)$ for User 2 at time slot 1. At time slot 2, we let User 2 transmit along its best direction. In this case, we can choose the precoder

$$\mathbf{B}^2(4,4) = \frac{1}{\sqrt{2}}[\mathbf{U}_{G_2(4,4)}(4,4)(1), \mathbf{U}_{G_2(4,4)}(4,4)(1)] \tag{33}$$

For User 1 at time slot 2, at receiver one, $\mathbf{A}^2(4,4)$ needs to satisfy the following the following equation

$$(\mathbf{G}^2_{11}(4,4)(1))^{\dagger} \cdot \mathbf{H}^2_{11}(4,4)(1) = 0 \tag{34}$$

and the power constraint

$$||\mathbf{A}^2(4,4)(1)||^2 = \frac{1}{2} \tag{35}$$

By solving the above two equations, we can get the precoder $\mathbf{A}^2(4, 4)$ for User 1 at time slot 2. With our precoders $\mathbf{A}^1(4, 4)$, $\mathbf{B}^1(4, 4)$ at time slot 1 and $\mathbf{A}^2(4, 4)$, $\mathbf{B}^2(4, 4)$ at time slot 2, we can show that we can achieve interference-free transmission with low decoding complexity and full diversity simultaneously as shown in the next two sections.

IV   Decoding with Low Complexity

In the last section, we have shown how to build the orthogonal transmission structure. Once the orthogonal structure is built, it is easy to realize low-complexity decoding. In this section, we will show how to decode and analyze the decoding complexity. We first consider the decoding at receiver one. In Equation (22), if we let

$$\overline{\mathbf{H}}_1(4,4) = \begin{pmatrix} \mathbf{H}^1_{11}(4,4)(1) & \mathbf{H}^1_{11}(4,4)(1) \\ (\mathbf{H}^2_{11}(4,4)(1))^* & -(\mathbf{H}^2_{11}(4,4)(1))^* \end{pmatrix} \tag{36}$$

and

$$\overline{\mathbf{G}}_1(4,4) = \begin{pmatrix} \mathbf{G}_{11}^1(4,4)(1) & \mathbf{G}_{11}^1(4,4)(1) \\ (\mathbf{G}_{11}^2(4,4)(1))^* & -(\mathbf{G}_{11}^2(4,4)(1))^* \end{pmatrix} \tag{37}$$

then Equation (22) becomes

$$\mathbf{y}_1(4,1) = \overline{\mathbf{H}}_1(4,4) \cdot \begin{pmatrix} c_1 \\ c_2 \end{pmatrix} + \overline{\mathbf{G}}_1(4,4) \cdot \begin{pmatrix} s_1 \\ s_2 \end{pmatrix} + \mathbf{n}_1(4,1) \tag{38}$$

Note that at receiver one, $\begin{pmatrix} c_1 \\ c_2 \end{pmatrix}$ are the desired signal and $\begin{pmatrix} s_1 \\ s_2 \end{pmatrix}$ are the interference. We can cancel the interference by multiplying both sides of Equation (38) by matrix $\overline{\mathbf{H}}_1(4,4)^\dagger$. Then we get

$$\overline{\mathbf{H}}_1(4,4)^\dagger \cdot \mathbf{y}_1(4,1) = \overline{\mathbf{H}}_1(4,4)^\dagger \overline{\mathbf{H}}_1(4,4) \begin{pmatrix} c_1 \\ c_2 \end{pmatrix} + \overline{\mathbf{H}}_1(4,4)^\dagger \mathbf{n}_1(4,1) \tag{39}$$

Here we have canceled the interference because $\overline{\mathbf{H}}_1(4,4)^\dagger \overline{\mathbf{G}}_1(4,4) = 0$. In order to decode the symbols, we first multiply both sides of Equations (39) by matrix $(\overline{\mathbf{H}}_1(4,4)^\dagger \overline{\mathbf{H}}_1(4,4))^{-\frac{1}{2}}$ to whiten the noise, i.e.,

$$(\overline{\mathbf{H}}_1(4,4)^\dagger \overline{\mathbf{H}}_1(4,4))^{-\frac{1}{2}} \overline{\mathbf{H}}_1(4,4)^\dagger \cdot \mathbf{y}_1(4,1) =$$
$$(\overline{\mathbf{H}}_1(4,4)^\dagger \overline{\mathbf{H}}_1(4,4))^{\frac{1}{2}} \begin{pmatrix} c_1 \\ c_2 \end{pmatrix} +$$
$$(\overline{\mathbf{H}}_1(4,4)^\dagger \overline{\mathbf{H}}_1(4,4))^{-\frac{1}{2}} \overline{\mathbf{H}}_1(4,4)^\dagger \mathbf{n}_1(4,1) \tag{40}$$

Then we can detect (c_1, c_2) by

$$\widehat{c}_1, \widehat{c}_2 =$$
$$\arg\min_{c_1, c_2} \left\| (\overline{\mathbf{H}}_1(4,4)^\dagger \overline{\mathbf{H}}_1(4,4))^{-\frac{1}{2}} \overline{\mathbf{H}}_1(4,4)^\dagger \right.$$
$$\left. \times \mathbf{y}_1(4,1) - (\overline{\mathbf{H}}_1(4,4)^\dagger \overline{\mathbf{H}}_1(4,4))^{\frac{1}{2}} \begin{pmatrix} c_1 \\ c_2 \end{pmatrix} \right\|_F^2 \tag{41}$$

Further, note that

$$\overline{\mathbf{H}}_1(4,4)^\dagger \overline{\mathbf{H}}_1(4,4) =$$
$$\begin{pmatrix} \|\mathbf{H}_{11}^1(4,4)(1)\|^2 + \|\mathbf{H}_{11}^2(4,4)(1)\|^2 & \|\mathbf{H}_{11}^1(4,4)(1)\|^2 - \|\mathbf{H}_{11}^2(4,4)(1)\|^2 \\ \|\mathbf{H}_{11}^1(4,4)(1)\|^2 - \|\mathbf{H}_{11}^2(4,4)(1)\|^2 & \|\mathbf{H}_{11}^1(4,4)(1)\|^2 + \|\mathbf{H}_{11}^2(4,4)(1)\|^2 \end{pmatrix} \tag{42}$$

So, when QAM is used, Equation (40) is equivalent to the following two equations.

$$\text{Real}\{(\overline{\mathbf{H}}_1(4,4)^\dagger \overline{\mathbf{H}}_1(4,4))^{-\frac{1}{2}} \overline{\mathbf{H}}_1(4,4)^\dagger \cdot \mathbf{y}_1(4,1)\}$$
$$= (\overline{\mathbf{H}}_1(4,4)^\dagger \overline{\mathbf{H}}_1(4,4))^{\frac{1}{2}} \text{Real}\{\begin{pmatrix} c_1 \\ c_2 \end{pmatrix}\}$$
$$+ \text{Real}\{(\overline{\mathbf{H}}_1(4,4)^\dagger \overline{\mathbf{H}}_1(4,4))^{-\frac{1}{2}} \overline{\mathbf{H}}_1(4,4)^\dagger \mathbf{n}_1(4,1)\} \tag{43}$$

$$\text{Imag}\{(\overline{\mathbf{H}}_1(4,4)^\dagger \overline{\mathbf{H}}_1(4,4))^{-\frac{1}{2}} \overline{\mathbf{H}}_1(4,4)^\dagger \cdot \mathbf{y}_1(4,1)\}$$
$$= (\overline{\mathbf{H}}_1(4,4)^\dagger \overline{\mathbf{H}}_1(4,4))^{\frac{1}{2}} \text{Imag}\{\begin{pmatrix} c_1 \\ c_2 \end{pmatrix}\}$$
$$+ \text{Imag}\{(\overline{\mathbf{H}}_1(4,4)^\dagger \overline{\mathbf{H}}_1(4,4))^{-\frac{1}{2}} \overline{\mathbf{H}}_1(4,4)^\dagger \mathbf{n}_1(4,1)\} \tag{44}$$

So we can detect the real part and the imaginary part of c_1, c_2 separately as follows:

$$\text{Real}\{\widehat{c}_1, \widehat{c}_2\} = \arg \min_{\text{Real}\{c_1, c_2\}}$$

$$\left\| \text{Real}\{(\overline{\mathbf{H}}_1(4,4)^\dagger \overline{\mathbf{H}}_1(4,4))^{-\frac{1}{2}} \overline{\mathbf{H}}_1(4,4)^\dagger \cdot \mathbf{y}_1(4,1)\} \right.$$
$$\left. - (\overline{\mathbf{H}}_1(4,4)^\dagger \overline{\mathbf{H}}_1(4,4))^{\frac{1}{2}} \text{Real}\{\begin{pmatrix} c_1 \\ c_2 \end{pmatrix}\} \right\|_F^2 \tag{45}$$

$$\text{Imag}\{\widehat{c}_1, \widehat{c}_2\} = \arg \min_{\text{Imag}\{c_1, c_2\}}$$

$$\left\| \text{Imag}\{(\overline{\mathbf{H}}_1(4,4)^\dagger \overline{\mathbf{H}}_1(4,4))^{-\frac{1}{2}} \overline{\mathbf{H}}_1(4,4)^\dagger \cdot \mathbf{y}_1(4,1)\} \right.$$
$$\left. - (\overline{\mathbf{H}}_1(4,4)^\dagger \overline{\mathbf{H}}_1(4,4))^{\frac{1}{2}} \text{Imag}\{\begin{pmatrix} c_1 \\ c_2 \end{pmatrix}\} \right\|_F^2 \tag{46}$$

The decoding complexity is symbol-by-symbol. Similarly, we can detect s_1, s_2 with symbol-by-symbol complexity at receiver two.

V   Diversity Analysis

In this section, we show that our proposed scheme can achieve full diversity for each user. We only prove that at receiver 1, the diversity for $c1$, $c2$ from user 1 is full. The proof for s_1, s_2 at receiver two will be similar. First, the diversity is defined as

$$d = -\lim_{\rho \to \infty} \frac{\log P_e}{\log \rho} \tag{47}$$

where ρ denotes the SNR and P_e represents the probability of error. We let $\mathbf{e} = \begin{pmatrix} e_1 \\ e_2 \end{pmatrix} = \begin{pmatrix} c_1 \\ c_2 \end{pmatrix} - \begin{pmatrix} \widehat{c}_1 \\ \widehat{c}_2 \end{pmatrix}$ denote the error vector. Here we add a rotation matrix $\mathbf{R}$ on the transmitted codewords to improve the system performance. Based on Equation (40), the pairwise error probability (PEP) for c_1, c_2 can be written as [18]

$$P(\mathbf{c} \to \overline{\mathbf{c}} | \overline{\mathbf{H}}_1(4,4)) =$$

$$Q\left(\sqrt{\frac{\rho \| (\overline{\mathbf{H}}_1(4,4)^\dagger \overline{\mathbf{H}}_1(4,4))^{\frac{1}{2}} \mathbf{R}(4,4) \mathbf{e}(4,1) \|_F^2}{4}} \right) =$$

$$Q\left(\sqrt{\frac{\mathbf{e}(4,1)^\dagger \mathbf{R}(4,4)^\dagger \overline{\mathbf{H}}_1(4,4)^\dagger \overline{\mathbf{H}}_1(4,4) \mathbf{R}(4,4) \mathbf{e}(4,1)}{4/\rho}} \right)$$

$$\leq \exp \left(\frac{\mathbf{e}(4,1)^\dagger \mathbf{R}(4,4)^\dagger \overline{\mathbf{H}}_1(4,4)^\dagger \overline{\mathbf{H}}_1(4,4) \mathbf{R}(4,4) \mathbf{e}(4,1)}{-4} \right)$$

$$= \exp \left(-\frac{\rho \lambda}{4/\rho} \right) \tag{48}$$

where

$$\lambda = \|\mathbf{H}_{11}^1(4,4)(1)\|_F^2 |\widehat{e}_1 + \widehat{e}_2|^2 + \|\mathbf{H}_{11}^2(4,4)(1)\|_F^2 |\widehat{e}_1 - \widehat{e}_2|^2 \tag{49}$$

and

$$\widehat{e}(4, 1) = \begin{pmatrix} \widehat{e}_1 \\ \widehat{e}_2 \end{pmatrix} = \mathbf{R}(4, 4)e(4, 1) \tag{50}$$

Since

$$\|\mathbf{H}_{11}^1(4, 4)(1)\|_F^2 \geq \frac{\|\mathbf{H}_1(4, 4)\|_F^2}{2} \cdot \frac{1}{2} \tag{51}$$

Inequality (48) can be written as

$$P(\mathbf{c} \rightarrow \overline{\mathbf{c}}|\overline{\mathbf{H}}_1(4, 4)) \leq \exp\left(-\frac{\rho\lambda}{4}\right)$$
$$= \exp\left(-\frac{\rho\|\mathbf{H}_1(4, 4)\|_F^2|\widehat{e}_1 + \widehat{e}_2|^2}{16}\right) \tag{52}$$

Therefore, we have

$$P(\mathbf{c} \rightarrow \overline{\mathbf{c}}) = E[P(\mathbf{c} \rightarrow \overline{\mathbf{c}}|\overline{\mathbf{H}}_1(4, 4))]$$
$$= E\left[\exp\left(-\frac{\rho\|\mathbf{H}_1(4, 4)\|_F^2|\widehat{e}_1 + \widehat{e}_2|^2}{16}\right)\right]$$
$$= \frac{1}{\prod_{j=1}^4(1 + \frac{\rho\tau}{16})} \tag{53}$$

Where

$$\tau = |\widehat{e}_1 + \widehat{e}_2|^2 \tag{54}$$

At high SNR region, (53) can be written as

$$P(\mathbf{c} \rightarrow \overline{\mathbf{c}}) \leq \left(\frac{\rho\tau}{16}\right)^{-4} \tag{55}$$

So the diversity is 4, full diversity, as long as $\tau \neq 0$. Also the coding gain is affected by τ and we can choose rotation matrix $\mathbf{R}(4, 4)$ properly to maximize τ. The best choice for rotation matrix depends on the adopted constellation. Such an optimization is a straightforward optimization that has been discussed in many existing literature [19]. Similarly, we can prove that the diversity for other codewords is also full.

VI   Simulation Results

In this section, we provide simulation results to evaluate the performance of the proposed scheme. First, we assume there are 2 transmitters each with 4 transmit antennas and 2 receivers each with 4 antennas. Each user uses our proposed scheme to transmit Alamouti codes to its receiver. Figure 2 presents simulation results using QPSK. We compare the performance of our scheme with that of two other scenarios that can achieve interference cancellation. In the first scenario, we use TDMA and beamforming. That is, at each time slot, only one transmitter sends signals to one receiver using beamforming. 16-QAM is used to have the same bit-rate. In the second scenario, each user uses the multi-user detection (MUD) method to send its codewords. The results show that our proposed scheme can achieve full diversity and symbol rate one. Note that we combine the array processing and space-time coding to avoid symbol rate loss. This does not mean that we cannot change the bit rate. We can always adapt the bit rate by changing the constellation according to the channel condition. In comparison, the TDMA and beamforming method can achieve full diversity but the rate is one half. The MUD method can achieve full rate, but it cannot achieve full diversity. As shown in the figure, our scheme provides the best performance due to its high diversity and increased coding gain without any rate loss.

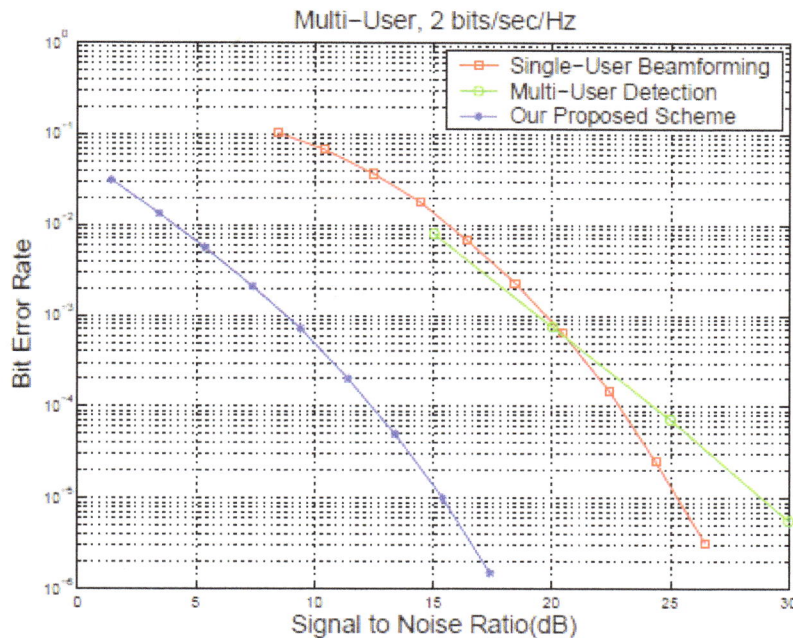

Figure 2: Simulation results for 2 users each with 4 transmit antennas and 2 receivers each with 4 receive antennas. The constellation is QPSK.

VII   Conclusions

In this paper, we propose an efficient transmission scheme for MIMO multi-user channels with two transmitters each with four transmit antennas and two receivers each with four receive antennas. By combining array processing and space-time coding, we can achieve full diversity and low decoding complexity. We analytically prove that our scheme can achieve low-complexity decoding and full diversity. Simulation results validate our theoretical analysis.

REFERENCES

[1] A. Zaier and R. Bouallegue, Blind Channel Estimation Enhancement for MIMO- OFDM Systems Under High Mobility Conditions, *International Journal of Wireless & Mobile Networks (IJWMN)*, 2012.

[2] F. Li, "Optimization of input covariance matrix for multi-antenna correlated channels," *International Journal of Computer Networks & Communications (IJCNC)*, 2011.

[3] D. C. Popescu, O. Popescu, and C. Rose, "Interference avoidance for multi access vector channels," in *Proc. Int. Symp. Inform. Theory*, July 2002, p. 499.

[4] J. Wang and K. Yao, "Multiuser spatio-temporal coding for wireless communications," in *Proc. IEEE Wireless Commun. Networking Conf.*, vol. 1, Mar. 2002, pp. 276-279.

[5] G. G. Raleigh and J.M. Cioffi, "Spatio-temporal coding for wireless communication," *IEEE Trans. Commun.*, vol. 46, pp. 357-366, Mar. 1998.

[6] H. Lu, R. Vehkalahti, C. Hollanti, J. Lahtonen, Y. Hong and E. Viterbo, "New space time code constructions for two-user multiple access channels," *IEEE J. Sel. Top. Sign. Proces.*, vol. 3, no. 6, pp.939-957, Dec. 2009.

[7] F. Li and H. Jafarkhani, "Space-Time Processing for X Channels Using Precoders," *IEEE Transactions on Signal Processing.*

[8] F. Li, *Multi-Antenna Multi-User Interference Cancellation and Detection Using Precoders*, PhD thesis, UC Irvine, 2012

[9] P. Wolniansky, G. Foschini, G. Golden, and R. Valenzuela, "V-BLAST: An architecture for realizing very high data rates over the rich scattering wireless channel," in *Proc. ISSSE*, Sept. 1998.

[10] G. J. Foschini and M. J. Gans, "On limits of wireless personal communications in a fading environment when using multiple antennas," *Wireless Pers. Commun.*, vol. 6, pp. 311-335, Mar. 1998.

[11] A. F. Naguib, N. Seshadri, and A. R. Calderbank, "Applications of space-time block codes and interference suppression for high capacity and high data rate wireless systems," in *Proc. Asilomar Conf. Signals, Systems and Computers*, 1998.

[12] F. Li and H. Jafarkhani, "Multiple-antenna interference cancellation and detection for two users using quantized feedback," *IEEE Transactions on Wireless Communication*, vol. 10, no. 1, pp. 154-163, Jan 2011.

[13] F. Li and H. Jafarkhani, "Interference cancellation and detection for multiple access channels with four users," in *Proceedings of IEEE International Conference on Communications(ICC 2010)*, June 2010.

[14] A. Stamoulis, N. Al-Dhahir and A. R. Calderbank, "Further results on interference cancellation and space-time block codes," in *Proc. 35th Asilomar conf. on Signals, Systems and Computers*, pp. 257-262, Oct. 2001.

[15] F. Li and H. Jafarkhani, "Interference Cancellation and Detection Using Precoders," *IEEE International Conference on Communications (ICC 2009)*, June 2009.

[16] S.Karmakar, M. K. Varanasi, "The diversity-multiplexing tradeoff of the MIMO Z interference channel," *Proc. IEEE int. Symp. on Inform. Theory*, pp 2188-2192, Jul. 2008.

[17] C. Sun, N.C. Karmakar, K.S. Lim, A. Feng, "Combining beamforming with Alamouti scheme for multiuser MIMO communications," in *Proceedings of Vehicular Technology Conference*, 2004.

[18] J. Huang, E. Au, and V. Lau, "Precoding of space-time block codes in multiuser MIMO channels with outdated channel state information," in *Proceedings of the IEEE International Symposium on Information Theory (ISIT '07)*, June 2007.

[19] F. Li, "Array Processing for Multi-User Multi-Antenna Interference Channels Using Precoders," *Wireless Personal Communications,* 2012.

[20] F. Li and H. Jafarkhani, "Using quantized feedback to cancel interference in multiple access channels," in *Proceedings of IEEE Global Telecommunications Conference(Globecom 2010)*, December, 2010.

[21] E. Malkam and H. Leib, "Coded diversity on block-fading channels," *IEEE Trans. Inf. Th.*, vol. 45, no. 2, Mar. 1999.

[22] L. Ozarow, S. Shamai, and A. D. Wyner, "Information theoretic considerations for cellular mobile radio," *IEEE Trans. Veh. Technol.*, vol. 43, pp. 359-378, May 1994.

[23] V. R. Cadambe and S. A. Jafar, "Interference alignment and the degrees of freedom for the K user interference channel," *IEEE Transactions on Information Theory*, vol. 54, no. 8, pp. 3425-3441, Aug 2008.

[24] F. Li and Q. T. Zhang, "Transmission strategy for MIMO correlated rayleigh fading channels with mutual coupling," in *Proceedings of IEEE International Conference on Communications (ICC 2007)*, June, 2007.

[25] F. Li and H. Jafarkhani, "Resource allocation algorithms with reduced complexity in MIMO multi-hop fading channels," in *Proceedings of IEEE Wireless Communications and Networking Conference*, 2009.

[26] S. M. Alamouti, "A simple transmit diversity technique for wireless communications," *IEEE J. Select. Areas Commun.*, vol. 16, no. 8, pp. 1451-1458, Oct. 1998.

[27] M. K. Simon and M.-S. Alouini, *Digital Communication over Fading Channels*, 1st ed. New York: Wiley, 2000.

[28] F. Li and H. Jafarkhani, "Multiple-antenna interference cancellation and detection for two users using precoders," *IEEE Journal of Selected Topics in Signal Processing*, December 2009.

[29] F. Li and H. Jafarkhani, "Interference Cancellation and Detection for More than Two Users," *IEEE Transactions on Communications*, March 2011.

[30] E. Bayer-Fluckiger, F. Oggier, and E. Viterbo, "New algebraic constructions of rotated *Zn*-lattice constellations for the Rayleigh fading channel," *IEEE Trans. Inform. Theory*, vol. 50, pp. 702-714, Apr. 2004.

4

A SIMPLE PRIORITY-BASED SCHEME FOR DELAY-SENSITIVE DATA TRANSMISSION OVER WIRELESS SENSOR NETWORKS

Farshad Safaei[1], Hamed Mahzoon[2], Mohammad Sadegh Talebi

[1] Faculty of ECE, Shahid Beheshti University G.C., Evin 1983963113, Tehran, IRAN
f_safaei@sbu.ac.ir
[2] Department of System Innovation Graduate School of Engineering Science
Osaka University, Osaka, Japan
hamed.mahzoon@irl.sys.es.osaka-u.ac.jp
[3] School of Computer Science Institute for Research in Fundamental Sciences (IPM)
Tehran, Iran
mstalebi@ipm.ir

ABSTRACT

In the course of last decade, wireless sensor networks (WSNs) have grabbed attention of both academic research community and industrial users. Such networks provide a broad range of applications, making them of great significance. Energy is the greatest concern of such battery-operated networks as it directly influences the lifetime of the network. As packet dropping due to congestion has a dramatically negative impact on energy consumption, congestion control is a vital issue for wireless sensor network, particularly for wireless multimedia sensor networks (WMSNs) in which bursty traffic is prevailing. In this paper, we present a simple yet efficient priority-aware congestion control scheme to support bursty data. Results achieved from our extensive simulation experiments corroborate that the proposed scheme operates well in terms of delay, throughput, and packet dropping, for both bursty and ordinary traffic.

KEYWORDS

Sensor Networks, Quality of Service, Congestion Control, Source Priority, Essential Queue

1. INTRODUCTION

A wireless sensor network consists of sensor nodes deployed over a geographic area mainly for monitoring physical phenomena like temperature, humidity, vibrations, seismic events, etc. [1]-[3]. Typically, a sensor node is a battery-operated device with a limited budget of energy, which consists of three basic components: a sensing element for data acquisition from environmental events, a processing subsystem for local data processing and storage, and a transmission device that provides wireless communication for nodes. As sensor nodes are usually deployed over environments that are almost unreachable, recharging the power sources proves cumbersome. This brings out the power consumption as a challenging issue and consequently, energy conservation is a key issue in designing WSNs [1].

Since WSN nodes are not always deployed over predetermined locations, the network needs to be designed with some levels of self-configuration capability. Once the nodes are deployed, the network is left unattended to perform monitoring, thus the network topology changes due to alternative node failures, suspensions, temporary droppings, and environmental distortion. The relatively higher number of nodes in WSNs makes its management much more complicated. A typical wireless sensor network is depicted in Figure 1, where nodes are deployed over the area referred to as sensor field. Each sensor node is capable of sensing the intended physical phenomena and routing the monitoring information toward the sink node through multipath with no infrastructure. The sink node is connected to a task manager node via Internet or satellite.

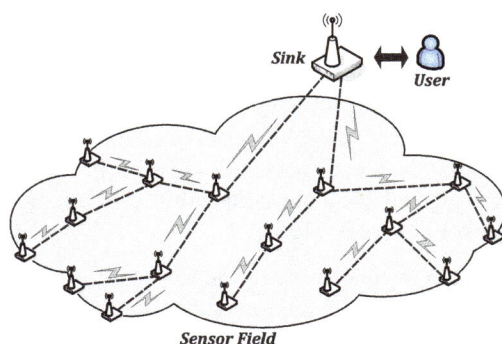

Figure 1. A wireless sensor network deployed in a sensor field

With advances in VLSI and production of MEMS, nowadays it's possible to integrate sensing, video/audio processing, and communication apparatus into a single tiny node at low prices. These advancements have been led to emergence of wireless multimedia sensor networks (WMSNs) which are missioned to sense, process, and transport image/audio/video streams both reliably and efficiently. A typical node in WMSNs is equipped with multimedia recording devices such as camera and microphones, by which two different types of data acquisition is collected: snapshot and streaming data. The first multimedia data type delivers event triggered observations obtained in a short period of time. On the contrary, the second multimedia content is generated over longer time periods. Multimedia applications typically publish great amount of data requiring high transmission rate and huge processing. Thus, compared to traditional WSNs, in such networks, the role of computation power is much more important. Characteristics that are common to sensor networks, such as resource constraints, unbalanced traffic and data redundancy, also exist in WMSNs [4]. We note, however, that for the case of WMSNs, the challenges due to these characteristics are often of greater concerns.

Supporting the preliminary requirements for providing Quality-of-Service (QoS), which is necessary for multimedia applications, is directly related to energy consumption, delay, reliability, distortion, and network lifetime. There is an inevitable correlation between levels of quality of accessible services in WMSNs and energy consumption in these networks, while obtaining any of these bases acquires the influential interaction on the other.

There exist several studies that tried to propose congestion control mechanisms that are aware of different packet priorities [5] and [6]. Such priority-aware schemes provide good performance to guarantee QoS for packets with different importance levels. However, they evince poor performance when they face bursty traffics. In this paper, we present a novel congestion control mechanism that aims to mitigate huge packet dropping in the presence of bursty traffic. It also tries to provide the required QoS for data of vital priority. To this end, the proposed mechanism leverages existing node architectures such as those proposed in [5] and [6], with only slight modifications. These modifications would only incur augmentation of a separate queue in each sensor node, thereafter referred to as essential queue. This essential queue is devoted to transfer delay-sensitive data over wireless sensor network. Thus, in the first place, all nodes send data buffered in their essential queues. If there is no packet in the essential queue, nodes will obey the rules dictated by priority-based rate adjustment schemes. We demonstrate QoS improvements asserted by extensive simulation experiments carried out in OMNET++.

The remainder of this paper is organized as follows. In the next section, a review of related works regarding QoS in WSNs is indicated. Then, a preliminary view to our scheme is mentioned in Section 3. Section 4 is devoted to our extensive experimental results. Finally, Section 5 concludes the paper.

2. RELATED WORK

This section is devoted to review some related studies. In the last few years, the networking research community has witnessed the emergence of a plethora of works concentrated on efficient communication protocol for wireless sensor networks. A large portion of such works have considered QoS guarantee in sensor network, e.g. [7]- [11]. Some other works concerned with providing light, efficient, and scalable congestion control protocols to possibly support several traffic classes in sensor networks [12]- [27].

In order to manage sudden transfer of data measured by sensor nodes, the authors of [13] have proposed CODA, as a congestion detection and avoidance armed with the following three mechanisms: receiver-based congestion detection; open-loop hop-by-hop back-pressure; and closed-loop multi-source regulation. He et al. [14] proposed SPEED, which is a real-time communication protocol supporting unicast, area-multicast, and area-anycast, realtime traffics. SPEED takes advantages such as being stateless, localized, scalable, and having minimal control overhead.

In [15], the authors presented a distributed and scalable algorithm to mitigate congestion occurred in a sensor network. The authors also aimed to guarantee fairness by trying to deliver equal number of packets by each node. In Yaghamaee et al. [6], a protocol is presented in which by utilizing queuing techniques and a novel weight nomination mechanism, either avoids probable network congestion, yet tries to esteem data priority. In this mechanism, for every child node of an individual node, a queue is assigned where the data transmitted from the child node flows through the corresponding queue. The data congestion evaluation is provided defining a congestion index. In [5], the authors proposed a novel congestion control protocol, dubbed Priority-based Congestion Control Protocol (PCCP), for wireless sensor networks. PCCP employs node priority index and congestion degree and enables cross-layer optimization and works under both single path and multipath routing. In [23], following Network Utility Maximization (NUM) framework [22], the authors aimed to formulate congestion control as the solution to an optimization-driven flow control problem with lifetime and link interference constraints. Similar to approaches used in NUM, the authors proposed a distributed and asynchronous flow control algorithm.

The authors of [17] have tried to improve network reliability by applying multipath data transmission. With this technique, there are multiple paths versus single shortest path to transmit data through, thus high density and congestion of data, containing multimedia records, on single shortest paths will not happen. This paper has also classified the data types and tried to perform more conscious node scheduling to moderate delay detriment. Finally, equipping nodes with multichannel transmission and exclusive transmission frequency provides a high throughput in this mechanism.

A typical sensor network has been simulated to evaluate lifetime and cost of the system by Cheng et al. [18]. In this paper, various cases are considered for different node variations like mobility, nodal structures, generated traffic types, and node energy allocation. In such cases, lifetime and cost of the network are evaluated and compared.

An evaluation model for performance of the sensor networks is presented by Chiasserini et al. [20], which compares the sensor networks with or without actuator and exhibits the differences. With the presented model, the network is simulated while the sensors are powered off. Once an event happens, the sensors are powered on via actuator nodes and start to record and transmit the phenomena.

Breaking down data into two parts, real-time and normal data, Akkaya et al. [21] innovated a technique in which real-time packets do not delay more than expected limit. Also, this mechanism tries to maximize throughput of normal packets. The nodes near to destination dynamically inform the far nodes about processing rate of real-time data so the network nodes alternatively adjust their processing rates. Calculation of approximate delay of real-time packets for different paths is led to an optimization problem, and hence, a path that minimizes the cost function is selected among different choices.

3. CONGESTION-AWARE ARCHITECTURE

In wireless multimedia sensor networks (WMSNs) different data types might be in progress to be transferred to the sink node. Most of such data are not delay-sensitive to which we refer to as ordinary data. Neither multiple transmissions of these data and dropping of excess ones impose any large data overhead to the network, nor is the received data density too high. Unlike such delay-insensitive data, others such as imaging and sonic data are almost delay-sensitive, and due to their inherent value, they are occasionally generated and transmitted in WMSNs, thus making them an essential data type. On the other hand, transmitting these data types through the network from the source node to the destination requires a great amount of energy from the nodes visited on the path. Thus, an overflow of such data in an intermediate node's queue imposes huge power waste to the sensor system. Regenerating and retransmitting lost data might add even more overhead to the network. These above mentioned challenges suggest that there should be more considerations about essential data and overhead associated to them.

It seems that classification of data can successfully handle the aforementioned challenges. We note, however, that providing this mechanism will exclusively cause loss of major amount of data types other than multimedia type in the network and thus, the network will change to an application specific one, in which the other data types can route the paths only if there is no multimedia data to be processed or transmitted. Though, managing data according to their types may cause mis-recognition of essential packets, since they might not be of multimedia ones. Meanwhile, there might be a very important such data that a user prefers to cancel receiving any other data to receive this data sooner. Thus, such data should take different priorities during the production, and in intermediate nodes different policies are pursued to apply these priorities.

In this section, we present our scheme to provide a simple yet efficient scheme for source-to-sink transmission of delay-sensitive data mentioned above. Our scheme employs the priority-based architecture proposed in [24]. However, as such a hierarchical scheme would not guarantee timeliness for delay-sensitive data delivery; we provision a separate queue, thereafter referred to as essential queue, for such data at each node. Figure 2 shows the architecture for our scheme.

In the sequel, we briefly review priority-based rate control scheme proposed in [5] and [24] that will be used for upstream rate allocation for ordinary data at each node.

3.1. Priority-Bases Rate Control

Assume that $T_s(i)$ and $\overline{T}_s(i)$ respectively show the service time of the current packet and the average service time in node i. The average service time is obtained by a moving average process. Then, let r_i denote the rate at which node i can transmit packets over the MAC layer. Then r_i is given by [24]:

$$r_i = \frac{1}{\overline{T}_s(i)} \tag{1}$$

Congestion can be quantified by congestion index according to the relevant queue size. Letting $q^k(i)$ represent the size of k-th queue in node i, the congestion index $I_X^k(i)$ is defined as a non-decreasing function of $q^k(i)$. In [24], the congestion index of k-th queue is defined according to a piecewise-linear model expressed by

$$I_X^k(i) = \begin{cases} \varepsilon & \text{if} \quad q^k(i) \le \min_{th}^k(i) \\ \dfrac{q^k(i) - \min_{th}^k(i)}{\max_{th}^k(i) - \min_{th}^k(i)} \max_p & \text{if} \quad \min_{th}^k(i) \le q^k(i) \le \max_{th}^k(i) \\ \max_p + \varepsilon & \text{if} \quad q^k(i) \ge \max_{th}^k(i) \end{cases} \tag{2}$$

where $\max_{th}^k(i)$ and $\min_{th}^k(i)$ denote the upper and lower thresholds for k-th queue of node i, respectively. Moreover, ε and $\max_p$ are constants less than 1.

Each child k at node i is in possession of a queue. Thereby taking into account the traffic generated by source i and N_i to be the number of children of node i, there will be N_i+1 queues at node i. In order to avoid congestion, each node i determines the rate for each of these internal queues based on the priority of each queue and its congestion level. Using $I_x^k(i)$ defined above, for each child k we define

$$(3)\ \overline{I}_x^k(i) = \frac{1}{N_i}\left(1 - \frac{I_x^k(i)}{\sum_{j=1}^{N_i+1} I_x^j(i)}\right); \qquad k = 1,\ldots,N_i+1.$$

Nodes are assumed as having different priorities. For each node i, we represent the source priority and the jth child's priority by $SP(i)$ and $TP(j)$, respectively. Then, the total priority of node i is the sum of its source priority and the priority of its children, i.e.

$$TP(i) = \sum_{j \in C(i)} TP(j) + SP(i) \tag{4}$$

where $C(i)$ is the set of children of node i. It's apparent from the above equation that for a leaf node l, i.e. $C(l) = \varnothing$, we have $TP(l) = SP(l)$. Based on the notion of total priority, in [24] the authors have defined the proportional priority factors for each node i as follows

$$p_i^1 = \frac{SP(i)}{TP(i)}, \tag{5}$$

$$p_i^k = \frac{TP(k)}{TP(i)}; \qquad k = 2,\ldots,N_i+1. \tag{6}$$

Using this proportional priority factors, each node i will be able to calculate the rate for each child k, i.e. r_i^k, as follows

$$r_i^k = \omega_i^k r_i, \tag{7}$$

where ω_i^k is a weight factor that combines the congestion index and priority of child k in the following way

$$\omega_i^k = \frac{p_i^k \overline{I}_x^k(i)}{\sum_{j=1}^{N_i+1} p_i^j \overline{I}_x^j(i)} \tag{8}$$

3.2. Essential Packets and the Queue Model

Figure 2 demonstrates our proposed architecture, where a queuing unit is provided for each child node. Data of each child node flow through the corresponding queue unit, unless it is an essential data. In this case, the *essential data* flow through the essential queue. Each queue unit is equipped with a scheduler of its own, which classifies entering data and applies the local timing according to the corresponding predefined policy. Data generated the sensor node are led to a separate queue unit that classifies all generated data on traffic generator and directs them through corresponding queues. If data generated by traffic generator are also of essential type, they will flow through essential queue and will be treated as an essential packet and scheduled by the timing strategy of this queue.

The scheduler in each node prescribes the packet processing rate as follows: If essential queue is empty, i.e. there is no essential packet, the scheduler determines rate for each child node according to rate control method presented in subsection 3.1. However, if essential queue is not empty, the scheduler sends all essential packets first, and then pays attention to other packets containing ordinary data.

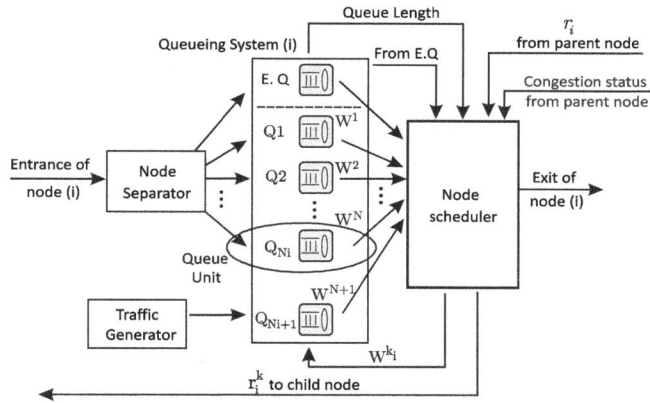

Figure 2. The proposed architecture

The priority-based rate control scheme described above assumes that child nodes generate data with the rate dictated by the parent node, and while the parent node detects crowd in a child node queue, it decreases this transmission rate to prevent dropping of data on that queue. However, in so many of WSNs, based on the usage, data might be generated in a bursty and regardless of nominated transmission rate, while most of them are essential ones. In this paper, we refer to such a deviation from nominated rate as norm-exceeding rate. If such norm-exceeding transmissions occur frequently, intermediate nodes will get in trouble since such norm-exceeding traffics are uncoordinated with the other data.

Uncoordinated data generation is referred to as data transmission with norm-exceeding rate. The first challenge is that such norm-exceeding traffics will cause congestion on the receiving buffer(s), while the ordinarily generated data fulfill the buffer, and once there is an empty room created in the buffer, again, a recently generated data occupies the available space. Thus, the later generated bursty data may find a full buffer and will be blocked. Even if the node treats these data as a high priority one, data overflow will be high over time, due to uncoordinated generation of such data types. The second challenge is that these data types are generated in a bursty and unpredictable fashion. Thus, once they are generated, the chance of dropping critical data as well as delayed adjustment of transmission rate for ordinary data due to high amount of such bursty data will increase dramatically.

In the next section, experimental results will show that how flowing such bursty data through essential queues would aim to dramatically alleviate packet dropping.

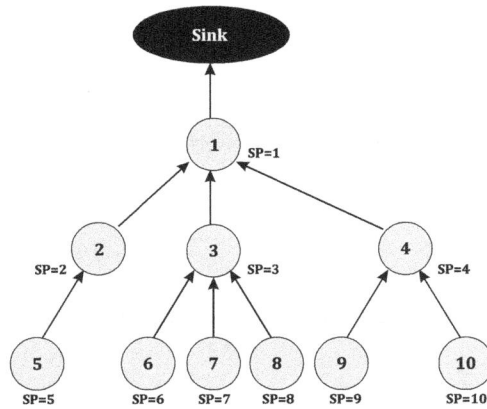

Figure 3. Topology for Simulation Experiments

4. EXPERIMENTAL RESULTS AND PERFORMANCE EVALUATION

In this section, we present the experimental results of the proposed architecture and compare the achieved results with those obtained from a system without using the proposed architecture. Performance of the essential queue has been explored through extensive experiments. The influence of norm-exceeding traffic on different aspects of network performance such as packet delay, packet dropping, and data throughput is also evaluated.

4.1. Simulation Setup

Simulation experiments are implemented in OMNET++ environment [28], which is a discrete event simulator. We carried out our experiments for the topology used in [24] which is a simple tree based topology shown in Figure 3. In this topology, the essential buffer length is 10 as well as the queue length of each child node. If the essential queue method is not used, the length of queues will grow up to reach the length of essential queue. If a node is equipped with the queue unit, buffers of all classes grow with the same amount of content. In the other word, unlike the case without using this scheme, when it is used the buffer for all the child nodes will grow with the same scale and this amount of growth will be equally divided among different classes in the queue unit.

4.2. Performance Assessment of Essential Queue

In this subsection, we examine utilizing the essential queue. Hence, we assume that all traffics generated by nodes are based on the feedback rate from the parent node. Thus, the essential data generation is also affected by the rate fed back from the parent node, and thus the rate of such essential data would diminish with the congestion detection and low feedback rate.

In a system without essential queue, it is expected that delay of essential packets exceeds that of a system with essential queue. On the other hand, it's expected that using this technique should not affect the ordinary packets and overall packet delay. These conjectures are validated by evaluating the packet delay simulation results which illustrate that the average packet delay for a system simulated without using the essential queue is 33 ms, and the packet delay through the simulation period varies from 20 to 60 ms. The same system is simulated with the utilization of essential queue. In this case, the packets delay varies between 2 to 70 ms and the average packet delay is about 33 ms, which is similar to the corresponding average in the system without essential queue. These results reveal that utilizing essential queue in a system, hopefully does not increase the packet delay of the system.

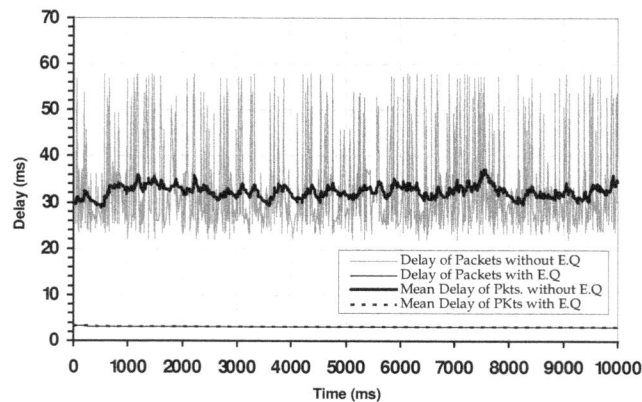

Figure 4. Essential packet delay of a system with and without utilizing essential queue

Now, we evaluate the delay of essential packets in above experiments. Figure 4 depicts the packet delay in our simulation. The curve indicated by "Without EQ" shows the system in which the essential queue is not used. The same result when the essential queue is provided in system is denoted in the curve indicated by "With EQ". As it was expected, due to lack of a special method for essential packets, this simulation also reveals the same result as the first case, meaning that the range of delay and the average delay similarly are the same as the first case. For "With EQ" curve, however, we can observe that the delay of essential packets, when the system is equipped with the essential queue, varies between 1 to 4.6 ms with the average equal to 3.2 ms. The results of this simulation extensively differ from the system in which the essential queue is not used. Considering ordinary packets, we obtain the results denoted in Figure 5. The results of systems without and with essential queue are illustrated and in Figure 5(a) and Figure 5(b), respectively.

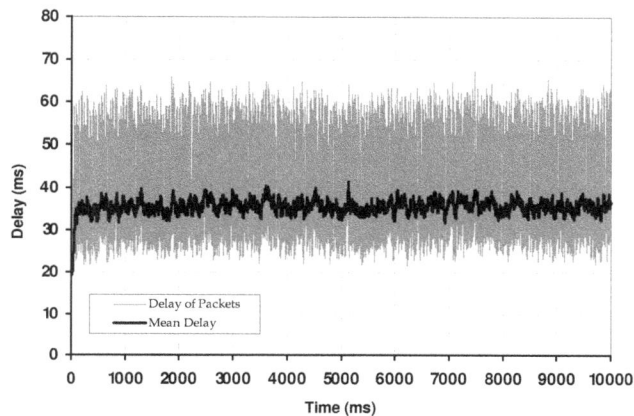

(a)

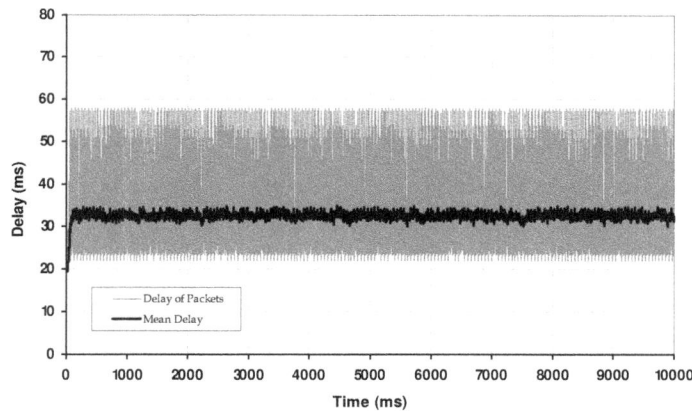

(b)

Figure 5. Delay of ordinary packets for a system (a) without utilizing essential queue, (b) with utilizing essential queue

Comparing these two figures, we realize that the average ordinary packet delay in the first system, i.e. 33 ms, is exacerbated up to 36 ms when the system is equipped like the second system, and this diminution is almost 9% of delay of ordinary packets. In congestion avoidance strategy, besides the status of intermediate queues, source priority should also be considered. It is expected that the source priority concept regulates the proportion of received data from different nodes. This issue is also improved using essential queue concept. Table 1 denotes the simulation results of the number of packets received by the last node of the network.

The first column of this table denotes the source priority while the second one shows the number of packets received from each priority. When there is no congestion and no bursty data received, it is expected to meet the source priority in this column. In the third column, the portion of received packets of each node to the received packets of base source is shown, while it depicts the actual priority observed in the network, and error of the calculated priority is given in the fourth column. This error is calculated based on the number of received packets of the base node, the number of received packets of the other nodes, the priority of the sources, the difference between the number of received packets, and expected number of received packets. Above evaluations have been applied to essential packets and the corresponding results are shown in columns 5 to 7. It is recognizable that due to the separate queue that has been provided for essential data, it is not expected to meet the priority of ordinary packets for essential data. Nevertheless, it seems that the essential packets are also received by the last node according to the different source priorities.

Table 1. Source priority and packet receipt with essential queue

Source Priority	Number of received packets	Proportion of received packet to base	Error (%)	Number of received essential packets	Proportion of received essential packets to base	Error (%)
1	181	1	0	17	1	0
2	362	2	0	39	2.29	14.7
3	544	3.01	0.18	65	3.82	27.5
4	725	4.05	0.13	66	3.88	-2.9
5	906	5.02	0.11	102	6	20
6	1088	6.01	0.18	101	5.94	-0.9
7	1271	7.03	0.32	127	7.47	6.7
8	1450	8.01	0.14	145	8.53	6.6
9	1633	9.03	0.25	170	10	11.1
10	1813	10.02	0.17	203	11.94	19.4

4.3. Performance of Essential Queue with Norm-Exceeding Traffic

Here, we evaluate the performance of essential queue concept while there is norm-exceeding traffic in the network. Thus, we assume that essential data are generated with no respect to nominated rate. Ordinary data are also generated with respect to the nominated rate and if there were essential data in the internal queue of a node, the essential data are transmitted and thereafter, the ordinary data are sent regarding the nominated rate. Figure 6 illustrates the mentioned traffic. Comparing "Without EQ" curve with the results of the first case discussed earlier, we see that the maximum packet delay is highly increased. Also, the average packet delay is about 500 ms, which is dramatically worse than the case with the traffic under the expected rate.

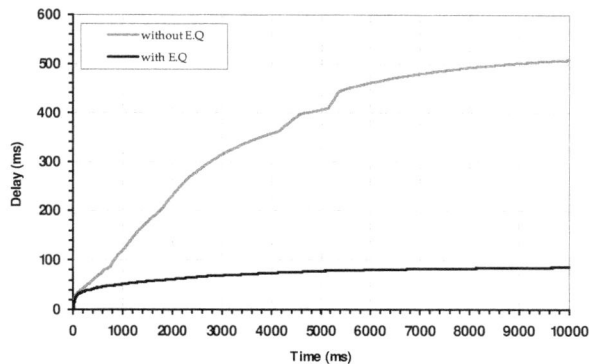

Figure 6. Packet delay of a system when norm-exceeding traffic is injected with and without utilizing essential queue

It is clearly depicted in the curve of the system utilizing essential queue that the average delay is approximately 90 ms, which is much better than the previous case. The reason for this great improvement stems from the way this system handles bursty traffics. When the essential queue is not used in the system, these essential data are generated in a bursty manner and flow through the corresponding child node. However, in the last system with the essential queue, the bursty data are directed to essential queue and they hierarchically traverse the network with essential queues toward the destination. Ordinary data then, realizing the bursty data due to the length of their queues, degrade the generating rate, and thus obviate the extra traffic. With this method the overall packet delay diminishes as well as packet dropping in the queues.

We accomplish the same evaluation on essential data and the results are portrayed in Figure 7. In the "With EQ" curve, the average packet delay is about 3 ms and the maximum delay is 6.3 ms. The significant point is that although the traffic rate is high and bursty, the packet delay is still preciously low, while the average delay is even lower than what is shown in "With EQ" curve of Figure 6. The average essential packet delay in the system without essential queue is about 900 ms with the maximum delay of 4832 ms. These results are extremely high compared with the system with the essential queue. The overall diagram is so similar to the corresponding system in Figure 6.

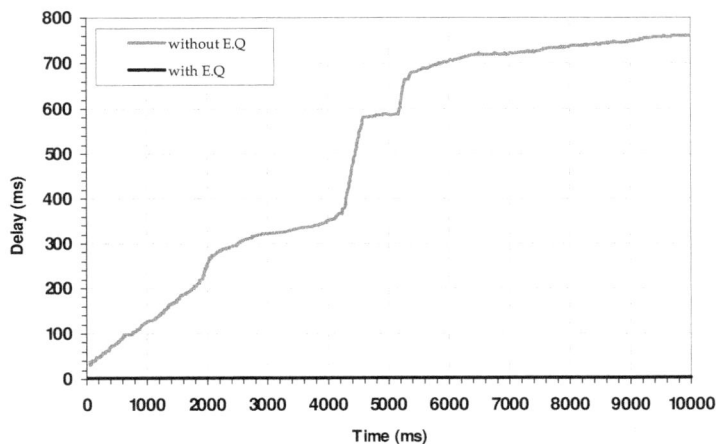

Figure 7. Essential packet delay of a system when norm-exceeding traffic is injected with and without utilizing essential queue

In Figure 8 we evaluate the results of above mentioned systems for ordinary data. It is expected that the overall shape of the diagram does not differ from the diagram for all packet types. Figure 8(a) depicts the result for the system not using the essential queue. As it seems, the average delay of ordinary packets is about 500 ms that conforms the all-type average delay. When using essential queue in this system, the results will be like Figure 8(b). The average packet delay in this case is 110 ms that is far better than that of not using essential queue. The reason of this improvement is that the ordinary data generation rate is not coordinated with the generation rate of bursty traffic. In the system without essential queue, although ordinary data justify their generation rate with the current queue status, because all the queues are crowded due to bursty traffic, the transmission rate adjustment is not appropriate regarding to the current network status. On the other hand, utilizing the essential queue in a system detaches the bursty traffic from ordinary data, and thus ordinary data are able to adjust their rate regarding the real network status. This issue will prevent from generation of traffics that sensor node is unable to support and handle.

In this part, the portion of received packets by the last node of the network is calculated. Table 2 summarizes the experiment in which essential queue has been used. Comparing these with the one without essential queue demonstrates that packet priority for essential data and ordinary data is improved by 10.9% and 4.9%, respectively. Figure 9(a)-9(b) comparatively depict the error of observing source priority in system with and without essential queue, respectively.

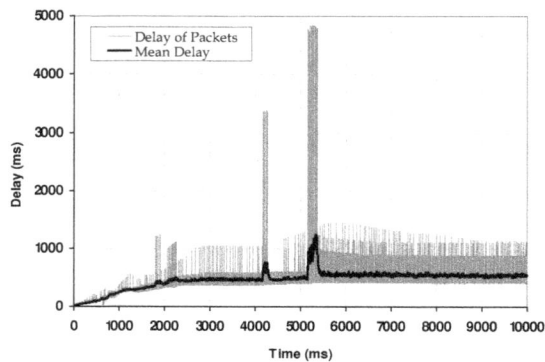

(a)

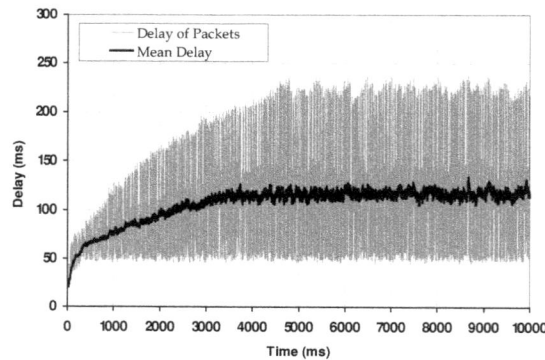

(b)

Figure 8. Delay of ordinary packets for a system with norm-exceeding traffic injection (a) without utilizing essential queue, (b) with utilizing essential queue

Table 2. Source priority and packet receiving with essential queue and norm-exceeding traffic

Source Priority	Number of received packets	Proportion of received packet to base	Error (%)	Number of received essential packets	Proportion of received essential packets to base	Error (%)
1	363	1	0	199	1	0
2	455	1.25	-37.3	198	0.99	-50.3
3	647	1.78	-40.6	214	1.07	-64.1
4	805	2.21	-44.6	211	1.06	-73.5
5	909	2.50	-49.9	208	1.05	-79.1
6	1068	2.94	-51.0	215	1.08	-81.9
7	1218	3.36	-52.1	205	1.03	-85.3
8	1319	3.63	-54.6	176	0.88	-88.9
9	1537	4.23	-52.9	208	1.04	-88.4
10	1656	4.56	-54.4	228	1.14	-88.5

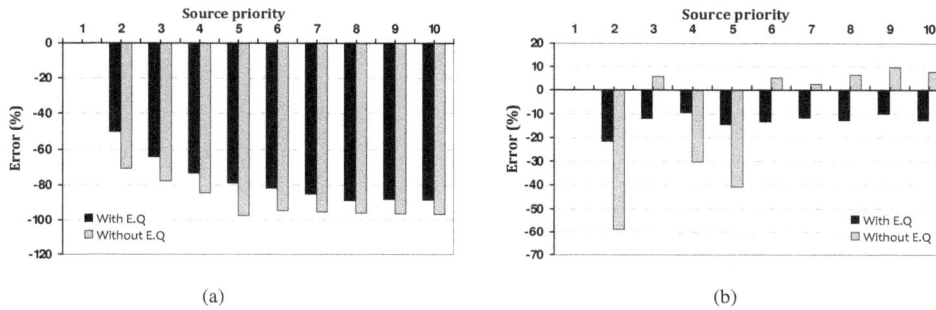

(a)

(b)

Figure 9. Source priority observation for norm-exceeding rate traffic: (a) for essential data, (b) for ordinary data

4.4. Packet Dropping

If all data transmission is done with the maximum rate proposed by the parent node, then packet dropping will not happen. On the contrary, when there are bursty and norm-exceeding essential data traffic in the network, packet dropping is inevitable. Here we show that how utilization of essential queue in a network alleviates packet dropping. First, we consider a system without essential queue and then we simulate the same system using essential queue and denote the data dropping diagram as Figure 10. In this figure, dropping percentage is increasing at first, but after reaching the steady state, data transmission is done by dropping of 21% of data. In this figure, data dropping starts after 700 ms, and after its beginning and its bursty growth, in the time period of 800 ms to 1200 ms, dropping is restrained. This is due to increased intermediate queue lengths and consequently growth of congestion index of the queues. Thus, nodes degrade the rate of their ordinary data. Hence, data dropping is controlled in the aforementioned time period. On the other hand, because of high rate of essential data, packet dropping percentage increases up to its steady state value.

This situation changes if the essential queue is utilized in the system. As it can be seen in Figure 10, after bursty growth of dropping percentage, from 2000 ms, the transient state of the system passes and dropping percentage reaches a constant rate. The reason for such a bursty dropping at the beginning of the traffic generation and transmission is the higher priority of essential data with respect to ordinary data. The existence of essential data prevents ordinary data to be serviced. While the nodes are calculating the generation rate of their child nodes, high generation rate of essential data causes dropping of ordinary data. After the required time, when the network is in the steady state, the generation rate of nodes is specified and dropping rate reaches its permanent value.

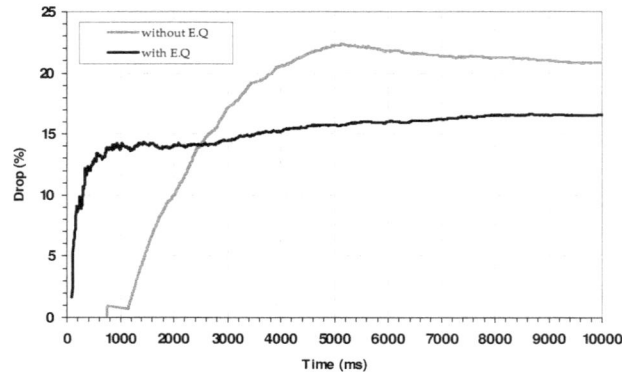

Figure 10. Packet dropping percentage when norm-exceeding traffic of parent node is used
In Figure 10, it has been shown that utilizing essential queue improves total data dropping by 5% thanks to usage of separated queues for bursty and ordinary data. This improvement is observed in the essential data (Figure 11).

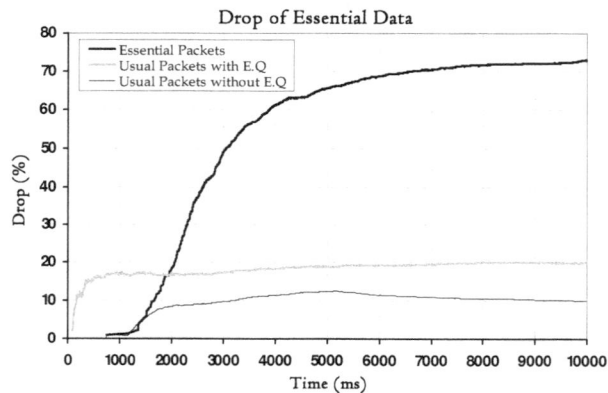

Figure 11. Packet dropping percentage when the parent traffic is norm-exceeding rate

In the system without essential queue, packet dropping rate is more than 70%. This high amount of dropping demonstrates that most of the essential data face a full queue after generation and competition for the corresponding queue. According to the coordination between the generation rate and the transmission rate of ordinary data, these data types mostly enter the queue when it is not full. Hence, utilizing essential queue in a system makes packet dropping as small as zero. It means that all droppings happened in Figure 10 for "with E.Q.", is not related to essential data and they are all ordinary data that have been dropped. Also, in Figure 11 two curves are depicted for ordinary data dropping while utilizing and not utilizing essential queue in the system. Comparing these two curves reveals that ordinary data dropping without using essential queue is about 10%. Then, ordinary data dropping starts to diminish. Hence, the comparison between these curves illustrates that there is 10% difference of dropping of ordinary packets, whether there is essential queue provided for the system or not. This means that we can exploit essential queue in a system, and with increasing ordinary data dropping from 10 to 20%, we can degrade the essential data dropping from 70% to 0, and total data dropping decreases from 21 to 16%, which admits the efficiency of utilization of essential queue.

4.5. Throughput

Now, we evaluate the performance of the network in terms of its throughput and investigate the influence of essential queue on this performance metric. We compare the essential packet throughput in a system with and without essential queue. We then study the ordinary data throughput of these systems. In Figure 12 the total data throughput is depicted while the system is equipped with and without the essential queue and norm-exceeding rate traffic is exerted. As it is shown, when the system doesn't use essential queue, in the time period of 500 to 5000 ms, throughput experiences an oscillatory behavior. Before 1000 ms, this value temporarily approaches zero. This is due to data dropping in the network because of buffer fulfillment and data generation rate adjustment by network nodes. In this figure, at the beginning, because of high packet dropping, throughput is deeply decreased and as it follows, due to adjustment of network rates to significant ones, the throughput approaches one. Finally, the average throughput in this case is about 0.9.

It is clear that from the beginning of the simulation, throughput is almost around 1 and as compared to Figure 13, here its oscillatory behavior is tolerable. From the beginning of the data transmission in the network, if there is essential data in a node, the ordinary data of that node are not sent and the corresponding queue will get a high amount of data. Thus, the congestion index of these queues will be high which will cause a quick adjustment of the generation rate of these data types. Hence, the rate adjustment of the ordinary data and even dropping of these data begin, and thus throughput will not be affected so much.

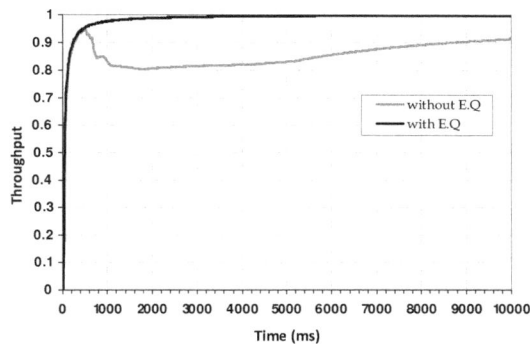

Figure 12. Data throughput for norm-exceeding rate traffic and in a system with and without essential queue

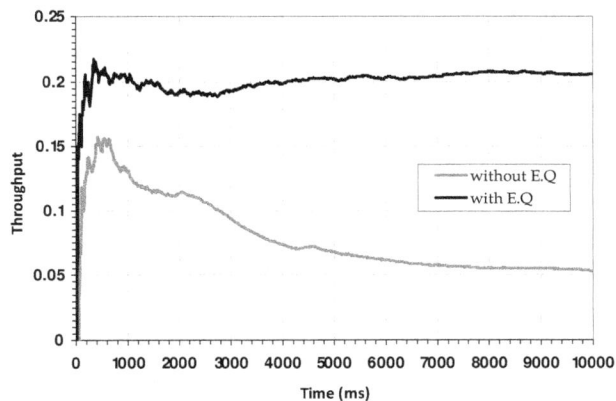

Figure 13. Essential data throughput for norm-exceeding rate traffic in a system with and without essential queue

The throughput of essential data is displayed in Figure 13. In the first figure, it is shown that although the throughput of essential data has higher values at the beginning of the simulation, but its average value is less than 0.1 and finally reaches the value 0.06 at the end.

Obviously, the throughput of essential data in such a system is increased. The steady state distribution of throughput during the simulation period illustrates that essential data transmission is done steadily. Comparing "With EQ" curves in both Figure 13 and Figure 12, we expect that ordinary data would also have high and almost constant throughput in this case. This expectation is validated with the results shown in Figure 14.

According to the throughput of total data in a system without essential queue, we expect that ordinary data throughput is also affected by the data whose generation rate is not coordinated with the rate proposed by the parent node. The result for such a system is displayed in Figure 14(a). As we expected, the ordinary data has the highest throughput too, which is validated due to non-existence of dropping that is seen in Figure 12. The throughput distribution is smooth with average of 0.8; thereby affection of bursty traffic is seen there. Thus, we conclude that utilizing essential queue not only has no negative effect on ordinary data, but also the process applied for essential data also improves the status of ordinary data in terms of dropping, delay, and throughput.

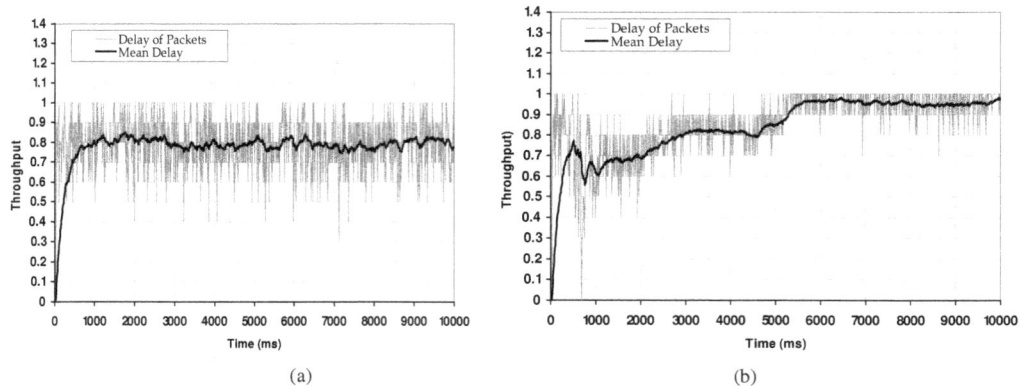

(a)   (b)

Figure 14. Ordinary data throughput for norm-exceeding rate traffic in a system (a) without essential queue, (b) with essential queues

In Figure 14(b) the above system is implemented without essential queue. As it was expected from the "Without EQ" curve of Figure 12 and Figure 13, in this case ordinary data are affected by the bursty generation of essential data and their throughput have been degraded precipitately. While, according to the "With EQ" curve of Figure 13, the throughput of essential data is of little variation, due to high amount of essential data generation and also dropping of ordinary data, their throughput catch extreme variations around the time period of 600 to 1400 ms. In the long run, when the network reaches steady state regime for data generation rate and dropping, ordinary data throughput tends to 86%.

5. CONCLUSION

In this paper we presented a novel priority based rate adjustment scheme for wireless sensor networks to manage the priority of delay-sensitive traffics in wireless sensor networks. Our scheme is based on the existing priority based schemes with an extra queue unit to serve delay-sensitive data that might be generated in a bursty manner. We introduced the essential queue to propose priorities for the packets and to recognize the essential data from others. Then, to evaluate the performance of the proposed scheme, we simulated an illustrative wireless sensor network with tree topology. Our experimental results showed that augmentation of such a simple queue unit to each node for supporting essential data, can greatly improve delay, throughput, and packet dropping rate of such delay-sensitive data. On the other hand, it would incur slight impact on such performance metrics for ordinary data which obey priority-based rate assigned by their parents.

REFERENCES

[1] I. F. Akyildiz, W. Su, Y. Sankarasubramaniam, and E. Cayirci (2002), "Wireless sensor networks: a survey," *Elsevier Computer Networks*, vol. 38, no. 4, pp. 393–422.

[2] J. Yick, B. Mukherjee, and D. Ghosal (2008), "Wireless sensor network survey," *Computer Networks*, vol. 52, no. 12, pp. 2292–2330.

[3] C. Y. Chong and S. P. Kumar (2003), "Sensor Networks: Evolution, Opportunities, and Challenges," *Proc. of the IEEE*, vol. 91, no. 8, pp. 1247–1256.

[4] I. F. Akyildiz, T. Melodia, and K. R. Chowdhury (2007), "A survey on wireless multimedia sensor networks," *Computer Networks*, vol. 51, no. 4, pp. 921–960.

[5] C. Wang, B. Li, K. Sohraby, M. Daneshmand, and Y. Hu (2007), "Upstream congestion control in wireless sensor networks through cross-layer optimization," *IEEE Journal on Selected Areas in Communications*, vol. 25, no. 4, pp. 786–795.

[6] M. H. Yaghmaee and D. Adjeroh (2008), "A New Priority Based Congestion Control Protocol for Wireless Multimedia Sensor Networks," *Proc. of WoWMoM 2008*, pp. 23-28.

[7] E. Felemban, C. -G. Lee, E. Ekici, R. Boder, and S. Vural (2005), "Probabilistic QoS Guarantee in Reliability and Timeliness Domains in Wireless Sensor Networks," *Proc. of IEEE INFOCOM 2005*, vol. 4, pp. 2646–2657.

[8] Y. Wu, S. Fahmy, and N. B. Shroff (2010), "Sleep/wake scheduling for multi-hop sensor networks: Nonconvexity and approximation algorithm," *Ad Hoc Networks*, vol. 8, no. 7, pp. 681–693.

[9] S. Tang and W. Li (2006), "QoS supporting and optimal energy allocation for a cluster based wireless sensor network," *Computer Communications*, vol. 29, no. 13–14, pp. 2569–2577.

[10] M. M. Alam, M. Mamun-Or-Rashid, and C. S. Hong (2006), "QoS-Aware Routing for Sensor Networks Using Distance-Based Proportional Delay Differentiation (DPDD)," *Int'l Conf. on Next-Generation Wireless Systems (ICNEWS)*, Dhaka, Bangladesh.

[11] H. O. Sanli, H. Cam, and X. Cheng (2004), "EQos: An Energy Efficient Qos Protocol for Wireless Sensor Networks," *Proc. of 2004 Western Simulation MultiConference*, San Diego, CA.

[12] S. Borasia and V. Raisinghani (2011), "A Review of Congestion Control Mechanisms for Wireless Sensor Networks," *Springer Journal of Technology Systems and Management*, pp. 201–206.

[13] C. Y. Wan, S. B. Eisenman, A. T. Campbell (2011), "Energy-efficient congestion detection and avoidance in sensor networks," *ACM Transactions on Sensor Networks (TOSN)*, vol. 7, no. 4, pp. 1–32.

[14] T. He, J. A. Stankovic, C. Lu, and T. Abdelzaher (2003), "SPEED: A Stateless Protocol for Real-Time Communication in Sensor Networks," *Proc. of IEEE International Conference on Distributed Computing Systems*, pp. 46–55.

[15] C. T. Ee and R. Bajcsy (2004), "Congestion control and fairness for many-to-one routing in sensor networks," *Proceedings of the 2nd international conference on Embedded networked sensor systems*, pp. 148–161.

[16] S. K. Dhurandher, S. Misra, H. Mittal, A. Agarwal, I.Woungang (2011), "Using ant-based agents for congestion control in ad-hoc wireless sensor networks," *Springer Cluster Computing*, vol. 14, no. 1, pp. 41–53.

[17] M. A. Hamid, M. M. Alam, and C. S. Hong (2008), "Design of a QoS-aware Routing Mechanism for Wireless Multimedia Sensor Networks," *IEEE GLOBECOM 2008*, New Orleans, LO, pp. 1–6

[18] Z. Cheng, M. Perillo, andW. B. Heinzelman (2008), "General Network Lifetime and Cost Models for Evaluating Sensor Network Deployment Strategies," *IEEE Transactions on Mobile Computing*, vol. 7, no. 4, pp. 484–497.

[19] I. Politis, M. Tsagkaropoulos, and S. Kotsopoulos (2008), "Optimizing Video Transmission over Wireless Multimedia Sensor Networks," *IEEE Global Telecommunications Conference (GLOBECOM)*, New Orleans, LO, pp. 1–6.

[20] C. F. Chiasserini and M. Garetto (2004), "Modeling the Performance ofWireless Sensor Networks," *Twenty-third Annual Joint Conference of the IEEE Computer and Communications Societies*.

[21] K. Akkaya and M. Younis (2003), "An Energy-Aware QoS Routing Protocol for Wireless Sensor Networks," *Proc. of 23rd International Conference on Distributed Computing Systems*, pp. 710-715.

[22] M. Chiang, S. Low, A. R. Calderbank, and J. C. Doyle, "Layering as Optimization Decomposition: A Mathematical Theory of Network Architectures," In Proc. IEEE, vol. 95, no. 1, pp. 255–312, Jan. 2007.

[23] J. Chen, W. Xu, S. He, Y. Sun, P. Thulasiraman, X. Shen (2010), "Utility-based asynchronous flow control algorithm for wireless sensor networks," *IEEE Journal on Selected Areas in Communications*, vol. 28, no. 7, pp. 1116–1126.

[24] M. H. Yaghmaee and D. A. Adjeroh (2008), "A Model for Differentiated Service Support inWireless Multimedia Sensor Networks," *Proc. of ICCCN 2008*, pp. 881–886.

[25] M. I. Khan, W. N. Gansterer, and G. Haring (2007), "Congestion Avoidance and Energy Efficient Routing Protocol for Wireless Sensor Networks with a Mobile Sink," *Journal of Networks*, vol. 2, no. 6, pp. 42–49.

[26] E. Toscano, O. Mirabella, and L. L. Bello (2007), "An Energy-efficient Real-Time Communication Framework for Wireless Sensor Networks," *Int'l Workshop on Real Time Networks (RTN)*, Pisa, Italy.

[27] F. Ren, T. He, S. Das, and C. Lin (2011), "Traffic-Aware Dynamic Routing to Alleviate Congestion in Wireless Sensor Networks," *IEEE Transactions on Parallel and Distributed Systems*, no. 99, pp. 1585–1599.

[28] A. Varga (2001), "The OMNeT++ discrete event simulation system," *Proceedings of the European Simulation Multiconference (ESM'01)*, pp. 319–324.

5

ANALYSIS OF A DETERMINISTIC JONG NANG GATE WITH TRANSMITTER COOPERATION

Moon Ho Lee[1], Md. Hashem Ali Khan[1] and Daechul Park[2]

[1]Division of Electronics & Information Engineering, Chonbuk National University, Jeonju 561-756, Korea
moonho@jbnu.ac.kr, hashem05ali@jbnu.ac.kr
[2] Dept. of Information & Communication Engineering, Hannam University, Daedeok-Gu, Daejeon 306-791, Korea
fia4joy@yahoo.co.kr

ABSTRACT

In this paper, we introduce practical the root of digital human binary coded Jong Nang communications as the wooden gate in Korea Jeju Island custom. We investigate Jong Nang gate models as an approximation of the AWGN model. The objective is to find a deterministic model, which is accessible to capacity analysis. Furthermore, this analysis should provide insights on the capacity of the AWGN model. Motivated by backhaul cooperation in cellular networks where cooperation is among base stations, we term the interference channel with conferencing transmitters. Jong Nang communications is normal 3 rafters placed on two vertical stones with three holes to convey the family's whereabouts that is deterministic signal, nowadays it is applied to backhaul in mobile base station.

KEYWORDS

Deterministic model, Jong Nang, Analysis of capacity & Gaussian Interference channel

1. INTRODUCTION

Jong Nang, the wooden gate in Korea Jeju Island dialect, had three wooden rafters placed on Jong-Ju-Mok (two large vertical stones with three holes) to convey the family's whereabouts. A product of the wisdom of Jeju Island people in Korea, the Jong Nang was a unique custom of local culture. As there was no gate at the house in Jeju Island, timbers were used to prevent cattle or horses from entering and having the barley and millet that were spread out in the yard. Later, Jong Nang was developed into the means of informing visitors whether the residents were at home or not [1-6].

(a) 3 gate open (0 0 0)   (b) 3 gate close (1 1 1)

*This paper was partially presented at the WiMoN Conference, 13-15 July 2012, Chennai, India.

The Jong Nang used the binary system similar to digital communications and computers today. Three timbers were exactly like three binary digits. The Jong Nang system could convey eight different messages. One of three Jong-Nang placed between the Jong-Ju-Mok, or "100" indicated there was no one at home, but the family would soon return form a neighboring area.

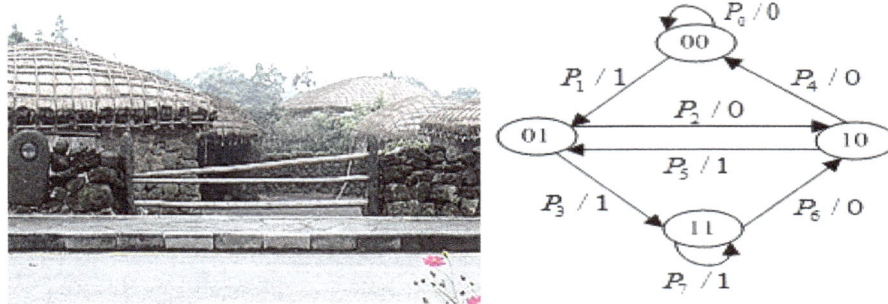

(c) 2 gate close: interference (0 1 1)   (d) Jong Nang Markov process

Figure 1. Korea Jeju Jong Nang

Two Jong Nangs or "101" meant the family was visiting a neighboring town and it would be a while before they returned. All these Jong Nangs or "111" announced the family was out of town for a long time, as shown in Fig 1. When none of the Jong Nang was placed, or "000", this meant the family was at home, as shown in Fig 1. This system derived from the life of the Jeju Island people. Table 1 shows the different Jong Nang messages. In paper [6] analysis of Jong Nang is one kind of practical deterministic scheme. Tse et al, Gamal & other person have written many papers as deterministic model with Gaussian interference channel [9-14].

Table 1. The comparison of Decimal and Binary number in Jong Nang

Decimal	Binary	Logic	Comparison
0	000		Staying at home
1	001		Visiting next door
2	010		Visiting next door
3	011		Visiting neighbour
4	100		Visiting next door
5	101		Visiting neighbour
6	110		Visiting neighbour
7	111		Long time out of home

From Table 1, binary MSB (Most significant bit) are $00,00,01,01,10,10,11,11$ and LSB (Least significant bit) are $0, 1, 0, 1, 0, 1, 0, 1$. The probability $(p_0 \cdots p_7)$ of Table 2 is similar as Fig 1(d). This process is Jong Nang Markov chains. Table 2 shows Markov state of the Jong Nang 3-NOR gate based on Table 1. The daily life of householder is staying or out of house based on Table 1.

Table 2. Markov state of Jong Nang 3-input NOR gate

Decimal	Binary	Probability	Decimal	Binary	Probability
0	000	P_0	4	100	P_4
1	001	P_1	5	101	P_5

| 2 | 010 | P_2 | 6 | 110 | P_6 |
| 3 | 011 | P_3 | 7 | 111 | P_7 |

The main motivation of this paper based on [6, 9, 10] AWGN capacity by Jong Nang gate modelling in Figure 2. The Jong Nang gate (JNG) is practical system and comparison Gaussian interference channel (IC).The JNG network is exact analysed and we get approximate analysis results. The deterministic model is defined in a way that allows for exact analysis of its capacity as Han-Kobayashi.

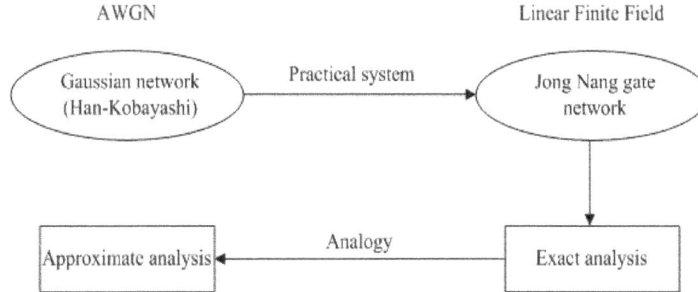

Figure 2. AWGN capacity by the Jong Nang gate modelling

The approximation of the Gaussian Interference channel (IC) by the q-ary expansion deterministic channel (QED) was first proposed by Avestimehr, Diggavi, and Tse. Bresler, Parekh, and Tse applied this approach to approximate the many-to-one Gaussian IC. The symmetric capacity achieving linear coding scheme for q-ary expansion deterministic IC is due to Jafar and Vishwanath. Bandermer showed that the entire capacity region can be achieved by this linear coding scheme. In this paper, we also introduce Jong Nang deterministic channel approach to transmitter's cooperation.

This paper is organized as follows. Section 2 describes practical system of Jong Nang gate model. Section 3 presents capacity of the Jong Nang channel as 3-input NOR channel. Section 4 presents the Jong Nang binary erasure multiple access channel. Section 5 JNG of Gaussian IC. Jong Nang gate interference channel with transmitter's cooperation and conclusions are delivered in Section 6 & 7.

2. PRACTICAL SYSTEM OF JONG NANG GATE MODEL

In this section, we introduce the practical system of JNG model in Figure 2. The idea deterministic modelling was may be most clearly explained by Tse [10]. The idea is to define a deterministic model in such a way that each AWGN network has a corresponding representation in the JNG model based on Table 1. The deterministic model is defined in a way that allows for exact analysis of its capacity. In the last step we hope to derive approximate results for the AWGN network by using the exact analysis from the deterministic network [9].

We might define any deterministic model and find the corresponding representation of our AWGN network. The two challenges are depicted in the lower half of the figure. Firstly, we have to analyse the capacity in the JNG model. Secondly, we have to show that the deterministic capacity is an approximation to the AWGN capacity in some sense.

Table 3. Backhaul

Technology	Max. Bandwidth	Min. node latency	Link latency	Max. length
Ethernet	100 Gbit/s	A few[3] μs	5 μ s/km (fiber)	40 km
AGPONI	10 Gbit/s shared	$\cdots 100 \mu$ s (upstr.)	5 μ s/km(fiber)	20 km
DVSL2	Mult[b].100Mbit/s	1.25...2ms[c]	<node latency	$\cdots 0.4$ km[d]
Microware	1 Gbit/s	$\cdots 100 \mu$ s	3 μ s/km (air)	A few[e] km

A very much simplified model will offer simple capacity analysis, but the approximation result might be difficult to find or might not hold, because the two models are rather unlike. On the other hand, a JNG model that is similar to the AWGN model might offer a shortcut to the approximation result, but finding its capacity might not be much simpler than the original problem. Now a day CoMP Backhaul is important for communications. We investigate limited transmitter or receiver cooperation helps mitigate interference. The behaviour of the benefit brought by transmitter cooperation is the same as by receiver cooperation. Backhauling for a CoMP enabled radio access network (RAN) therefore requires new backhaul technologies as shown in Table 3.

3. CAPACITY OF THE JONG NANG CHANNEL AS 3-INPUT NOR CHANNEL

When the Jong Nang channel is viewed as a coordinated 3-input noiseless multiple access NOR channel with the following transition probabilities: $p(y=1|x=100)= \quad p(y=0|x=100)= p(y=0|x=101)=p(y=0|x=111)=1$, the capacity of the channel can be calculated [7,8]

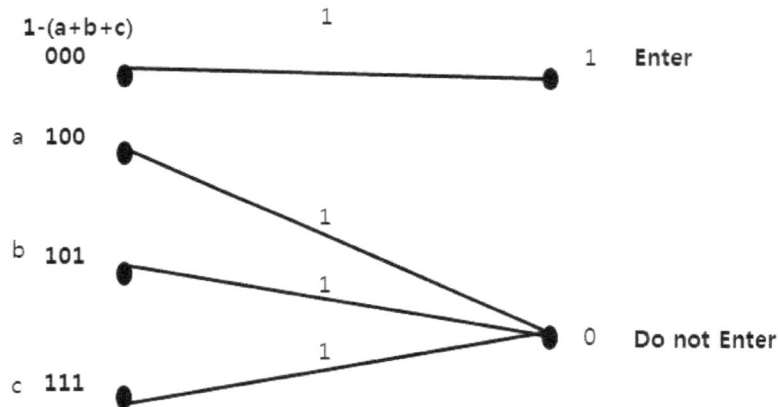

Figure 3. The Jong Nang channel

Theorem: The capacity of the Jong Nang Channel modelled as 3-input noiseless NOR channel is one.

Proof: Let $p(x=100)=a$, $p(x=101)=b$, $p(x=111)=c$, and $p(x=000)=1-(a+b+c)$. The output of the channel then has the following distribution:

$$p(y=0)=a+b+c, \quad p(y=1)=1-(a+b+c) \qquad (1)$$

Furthermore, we can calculate the joint distributions:

$$p(y=1,x=000)=p(y=1|x=000)p(x=000)=1-(a+b+c)$$

$$p(y=0, x=100) = p(y=0 \mid x=100) \, p(x=100) = a$$

$$p(y=0, x=101) = p(y=0 \mid x=101) \, p(x=101) = b$$

$$p(y=0, x=111) = p(y=0 \mid x=111) \, p(x=111) = c$$

Now, to compute the mutual information between the input and the output of the channel $I(X;Y) = H(Y) - H(Y \mid X)$, we evaluate $H(Y)$ and $H(Y \mid X)$. Let $q = a + b + c$, we have

$$H(Y) = -(1-q)\log(1-q) - q\log(q) = h(q) \tag{3}$$

$$H(Y \mid X) = -(1-q)\log(1) - a\log(1) - b\log(1) - c\log(1) = 0 \tag{4}$$

Hence,

$$I(X;Y) = H(Y) - H(Y \mid X) = h(q) \tag{5}$$

Maximizing the mutual information with respect to q,

$$\ln(2) \, dI(X;Y)/dq = 1 + \ln(1-q) - 1 - \ln(q) = \ln((1-q)/q) = 0, \tag{6}$$

which implies that $q = 1/2$.

Therefore, the capacity of the channel is $h(1/2) = 1$. **QED.**

That the capacity of the channel is 1 and can be achieved when $q = 1/2$ implies that the family can communicate at most one bit of information per three timbers using this channel, with the requirement that the family is home half of the time.

4. THE JONG NANG BINARY ERASURE MULTIPLE ACCESS CHANNEL

We can see that the Jong Nang code does not have any error. In the case that the channel has noise, which may result from a timber falling down from its place by natural or manmade sort of event, an error detecting code maybe desired. As shown Figure 3~ 4:

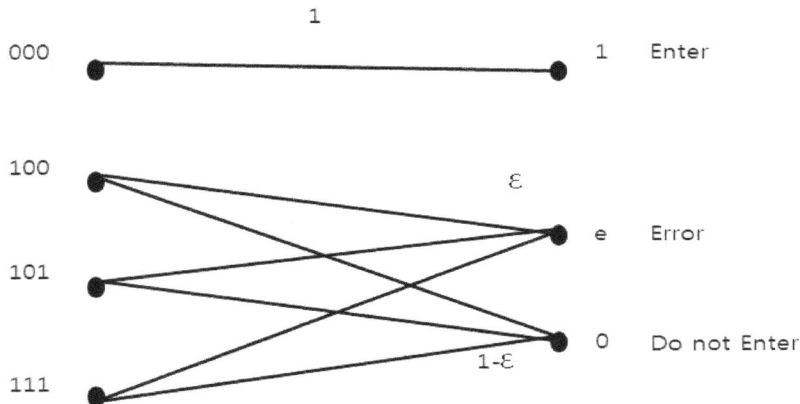

Figure 4. The Jong Nang binary erasure channel

As we mentioned, the channel does not take care the error number. It only takes whether the error is occurred or not. So all errors can be detected by using the codes, but the number of errors cannot be detected. In the following analysis, we assume that errors can occur when timbers fall down from its original place.

As shown Fig 5, the occured error includes all case of the errors; first, we transmit '111' which has the message 'out of town' and noise also occured, and we receive '101' which has the message 'visiting neighboring Village', this channel is transmitting the wrong message by ocuuring the error. Second, we transmit '111' which has the defined message and noise also occured. We receive '110' which has the undefined message.

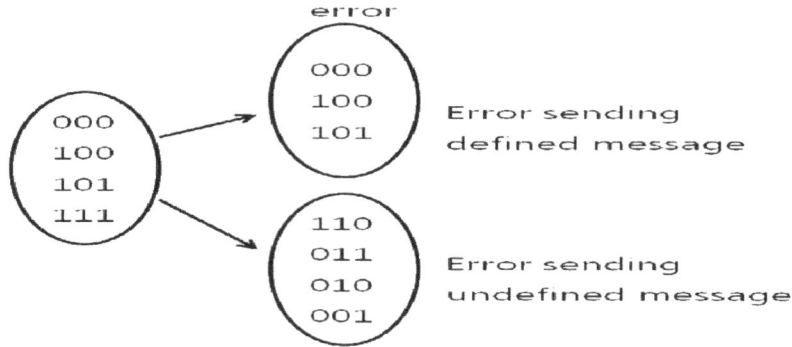

Figure 5. The event diagram of the occured error

Consequently, the only allowed words are 000, 011, 101, 110, 001, 010, 100. Furthermore, to reflect the situation in which the error is detect, the "erased" channel output is created, and let probabilities that an erasure occurs given channel inputs 000, 100, 101 and 111 are the same and equal to ε. That is when a visitor sees 110, 001, 010 or 001, channel output is set to "erased." Under these conditions, the channel transition probabilities are given,

$$p(y=1 \mid x=000)=1$$

$$p(y=0 \mid x=011)=p(y=0 \mid x=101)=p(y=0 \mid x=110)=1-\varepsilon$$

$$p(y=e \mid x=011)=p(y=e \mid x=101)=p(y=e \mid x=110)=\varepsilon \qquad (7)$$

Theorem: The capacity of the Noisy Jong Nang Channel as described above is 1.

Proof: Let $p(x=000)=a$, $p(x=100)=b$, $p(x=101)=c$, $p(x=111)=1-(a+b+c)$, and $q=a+b+c$. The output of the channel then has the following distribution:

$$p(y=1)=1-(a+b+c)=1-q,\ p(y=e)=\varepsilon(a+b+c)=\varepsilon q$$

$$p(y=0)=(1-\varepsilon)(a+b+c)=(1-\varepsilon)q \qquad (8)$$

Furthermore, we can calculate the joint distributions:

$$p(y=1,x=000)=p(y=1 \mid x=000)p(x=000)=1-(a+b+c)$$

$$p(y=0,x=100)=p(y=0 \mid x=100)p(x=100)=a(1-\varepsilon)$$

$$p(y=0,x=101)=p(y=0 \mid x=101)p(x=101)=b(1-\varepsilon)$$

$$p(y=0,x=111)=p(y=0|x=111)p(x=111)=c(1-\varepsilon)$$

$$p(y=e,x=100)=p(y=e|x=100)p(x=100)=a\varepsilon$$

$$p(y=e,x=101)=p(y=e|x=101)p(x=101)=b\varepsilon$$

$$p(y=e,x=111)=p(y=e|x=111)p(x=111)=c\varepsilon \qquad (9)$$

Now, to compute the mutual information between the input and the output of the channel $I(X;Y)=H(Y)-H(Y|X)$, we evaluate $H(Y)$ and $H(Y|X)$.

The capacity of the channel (C) is maximum of the mutual information; that is, when $H(Y|X)=0$, we find the capacity of the channel.

Let $q=a+b+c$, we have

$$H(Y)=-(1-q)\log(1-q)-\varepsilon q\log \varepsilon q-(1-\varepsilon)q\log(1-\varepsilon)q \qquad (10)$$

$$H(Y|X)=-(1-q)\log(1)-\varepsilon q\log \varepsilon-(1-\varepsilon)q\log(1-\varepsilon) \qquad (11)$$

Hence,

$$I(X;Y)=H(Y)-H(Y|X)=-(1-q)\log(1-q)-\varepsilon q\log \varepsilon q-(1-\varepsilon)q\log(1-\varepsilon)q$$

$$+(1-q)\log(1)+\varepsilon q\log \varepsilon+(1-\varepsilon)q\log(1-\varepsilon) \qquad (12)$$

Maximizing the mutual information with respect to q , $\ln(2)dI(X;Y)/dq$

$$=1+\ln(1-q)-\varepsilon-\varepsilon\ln \varepsilon q-(1-\varepsilon)-(1-\varepsilon)\ln(1-\varepsilon)q+\varepsilon+\ln \varepsilon+(1-\varepsilon)\ln(1-\varepsilon)$$

$$=\ln(1-q)-\varepsilon\ln q-(1-\varepsilon)\ln q=n(1-q)-\ln q=\ln[(1-q)/q]=0 \qquad (13)$$

which implies that $q=1/2$.

Therefore, the capacity of the channel is

$$C=-0.5\log(0.5)-0.5\varepsilon\log 0.5\varepsilon-0.5(1-\varepsilon)\log 0.5(1-\varepsilon)+0.5\varepsilon\log \varepsilon+0.5(1-\varepsilon)$$

$$\log(1-\varepsilon)=-0.5\log 0.5+0.5\varepsilon\log 2+0.5(1-\varepsilon)\log 2=0.5+0.5=1 . \ \textbf{QED.} \qquad (14)$$

We can analysis capacity of the Jong Nang Binary Erasure multiple access channel (MAC), in which four senders send the information to 3 receivers. This channel has binary inputs, $X_1,X_2 \in \{000,100,101,111\}$ and a ternary output Y_e,Y_1,Y_0.

Proof: Note that $p(x=000)=p(x=100)=p(x=101)=p(x=111)=1/4$. We can represent 000, 101, 110 and 111 by 00, 10, 01, 11 respectively. For the give information $X_1,X_2 \in \{000,100,101,111\}$ and ternary outputs $Y \in \{1,e,0\}$. We can get the channel capacity

$$C_1 < I(X_1;Y|X_2), \quad C_2 < I(X_2;Y|X_1)$$

We also get

$$C_1 = C_2 = Max\{I(X_1;Y|X_2)\}$$

$$= \sum_{X_1} \sum_{X_2} \sum_{Y} p(X_1) p(X_2) p(Y|X_1,X_2) \log \frac{p(Y|X_1,X_2)}{\sum_{X_1} p(X_1) p(Y|X_1,X_2)} \qquad (15)$$

$= 1$, as shown in Figure 6(a).

Then the channel capacity of the combined channel is

$$C_{12} = Max\, I(X_1,X_2;Y)$$

$$= \sum_{X_1} \sum_{X_2} \sum_{Y} p(X_1) p(X_2) p(Y|X_1,X_2) \log \frac{p(Y|X_1,X_2)}{\sum_{X_1} \sum_{X_2} p(X_1) p(X_2) p(Y|X_1,X_2)} \qquad (16)$$

It is easy to see that

$$P(Y=1) = P(X=000) = K,\; P(Y=\varepsilon) = P(X=100)\varepsilon + P(X=101)\varepsilon + P(X=111)\varepsilon$$

$$= \left[P(X=000) + P(X=101) + P(X=111) \right]\varepsilon = 3K\varepsilon,\text{ and}$$

$$P(Y=0) = \left[P(X=000) + P(X=101) + P(X=111) \right](1-\varepsilon) = 3K(1-\varepsilon) \qquad (17)$$

Consequently,

$$H(Y) = -\left[K\log K + 3K\varepsilon \log 3K\varepsilon + 3K(1-\varepsilon)\log K(1-\varepsilon) \right]. \qquad (18)$$

Since $K = 1/4$, $\varepsilon = 1/2$ and $H(Y|X_1,X_2) = 0$, we have

$$H(Y) = -\left[\frac{1}{4}\log\frac{1}{4} + 3\times\frac{1}{4}\times\frac{1}{2}\log\frac{3}{8} + 3\times\frac{1}{4}\times\frac{1}{2}\log\frac{3}{8} \right] = 1.56 \qquad (19)$$

and $Max(C_{12}) \approx 1.56$, which is shown in Figure 6(b).

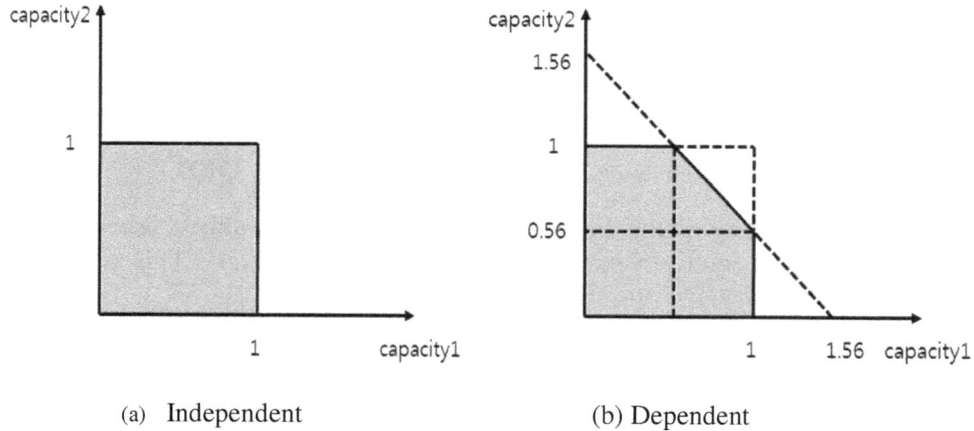

(a) Independent   (b) Dependent

Figure 6. Capacity region for the Jong Nang channels

In case Figure 6(a), the inputs of Jong Nang Erasure multiple access channel are independent, there is no interference between the senders. In case Figure 6(b), the inputs of Jong Nang channel are dependent, there is interference between the senders.

5. Jong Nang Gate of Gaussian IC

We introduce the q-ary expansion JNG interference channel and it closely the Gaussian IC in the limit of high SNR. Consider the symmetric case where the interference is specified by the parameter $\alpha \in [0, 2]$ and Table 1. Figure 7 shows JNG & Gaussian interference channel. We express X_1 as $\left[x_{1,L-1,}\, x_{1,L-2,}\, x_{1,L-3,} \cdots x_{1,0} \right]$ and similarly for X_2.

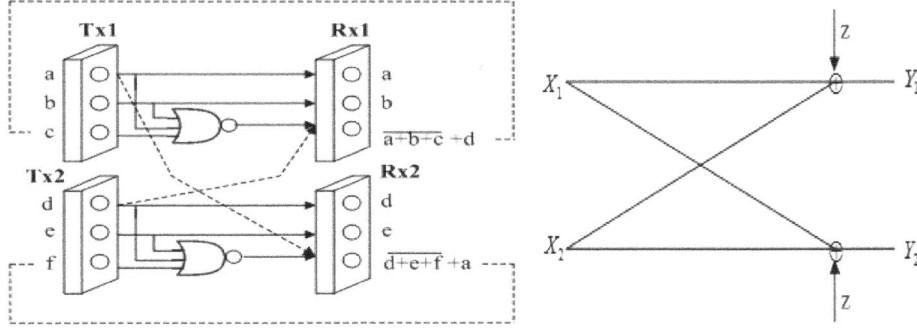

Figure 7. Block diagram of (a) JNG & (b) Gaussian IC

The JNG rafter up and down the output of the channel depends on whether the shift is positive or negative. The shift parameter is defined by

$$s = (\alpha - 1) L, \tag{20}$$

where L is q ary digit number. Now we get shift parameter from Eq. (1),

$$s = (\alpha - 1) L = (1/3 - 1)6 = -4.$$

where $\alpha = 1/3$, $L = 6$ from Fig 7(a).

5.1 Downshift

Here $0 \le \alpha \le 1$, Y_1 is a q-ary

$$Y_1 = \begin{cases} a \\ \overline{a + b + c} + d \pmod{q} \end{cases} \tag{21}$$

The outputs of the channel is represented as

$$Y_1 = X_1 + G_s X_2 \tag{22}$$

where G_s is a shift matrix.

5.2 Upshift

Here $1 \le \alpha \le 2$,

$$Y_2 = \begin{cases} d \\ \overline{d + e + f} + a \pmod{q} \\ c \end{cases} \tag{23}$$

The outputs of the channel is represented as

$$Y_2 = G_s X_1 + X_2 , \tag{24}$$

The capacity region of the symmetric QED-IC can be obtained by a straightforward of the capacity of the injective deterministic IC. The normalized capacity region is the set of rate pairs (R_1, R_2) such that

$$
\begin{aligned}
&R_1 \le 1, \ R_2 \le 1, \\
&R_1 + R_2 \le \max\{2\alpha, 2 - 2\alpha\} \\
&R_1 + R_2 \le \max\{\alpha, \ 2 - \alpha\}
\end{aligned}
\tag{25}
$$

for $\alpha \in [1/2, 1]$, and

$$
\begin{aligned}
&R_1 \le 1, \ R_2 \le 1 \\
&R_1 + R_2 \le \max\{2\alpha, 2 - 2\alpha\} \\
&R_1 + R_2 \le \max\{\alpha, 2 - \alpha\}
\end{aligned}
\tag{26}
$$

for $\alpha \in [0, 1/2]$.

The capacity region of the symmetric QED-IC can be achieved error free using a single linear coding scheme. For encoding, we use the matrix A. We get $[A]_6$ from Figure 7(a). Consider a binary expansion deterministic IC with $q = 2, L = 6$, and $\alpha = 1/3$.

$$
[A]_6 =
\begin{bmatrix}
1 & 0 & 0 & 0 & 0 & 1 \\
0 & 1 & 0 & 0 & 0 & 0 \\
0 & 0 & 1 & 0 & 0 & 0 \\
0 & 0 & 1 & 1 & 0 & 0 \\
0 & 0 & 0 & 0 & 1 & 0 \\
0 & 0 & 0 & 0 & 0 & 1
\end{bmatrix}
\bmod 2,
\tag{27}
$$

This decoding procedure corresponds to multiplying the output by the matrix B.

$$
[B]_6 =
\begin{bmatrix}
1 & 0 & 0 & 0 & 0 & -1 \\
0 & 1 & 0 & 0 & 0 & 0 \\
0 & 0 & 1 & 0 & 0 & 0 \\
0 & 0 & -1 & 1 & 0 & 0 \\
0 & 0 & 0 & 0 & 1 & 0 \\
0 & 0 & 0 & 0 & 0 & 1
\end{bmatrix}.
\tag{28}
$$

Note that $[B]_6 [A]_6 = [I]_6$ and $BG_s A = 0$, hence interference is cancelled while the intended signal is recovered perfectly. The transmits signal, $X_j = AU_j$, where A is q-ary matrix and U_j is uniformly distributed set of binary vectors. Decoder j multiplies its received symbol Y_j by a corresponding matrix B to recover U_j perfectly. For decoding case, $Y_j = BU_j$.

For encoding case,

$$X_1 = AU_1, \ X_2 = AU_2 \tag{29}$$

and decoding case,

$$Y_1 = BU_1, \ Y_2 = BU_2 \tag{30}$$

The transmitted symbol X_j and the received vector Y_j are shown in Figure 8. Decoding for U_1 can be performed as follows:

$$U_{1,0} = Y_{1,0}, \ U_{1,1} = Y_{1,1}, U_{1,4} = Y_{1,4}, U_{1,5} = Y_{1,5}$$

$$U_{1,3} = Y_{1,3} \oplus Y_{1,4} = Y_{1,3} \oplus U_{2,5} \text{ and } U_{1,6} = Y_{1,6} \oplus Y_{1,1} = Y_{1,6} \oplus U_{2,2}$$

The symmetric capacity is expressed as

$$C_{sym} = H(U_j) = I(U_j; Y_j) = I(X_j; Y_j) \tag{31}$$

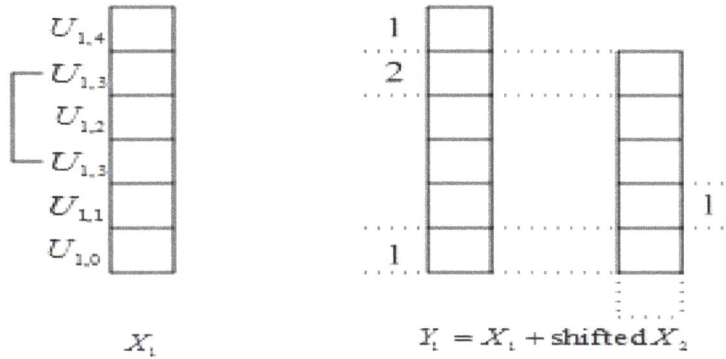

Figure 8. Transmitted symbol X_j and the received vector Y_j

6. JONG NANG GATE IC COOPERATION TRANSMITTERS

In this section, we discuss Jong Nang gate interference channel with cooperation transmitters, to overcome the complications both in achievability and inner bounds. The corresponding linear deterministic channel is parameterized by nonnegative integers $n_{11}, n_{21}, n_{22}, n_{12}, k_{12}$ and k_{21}, where

$$n_{ij} = \left(\left\lfloor \log |h_{ij}|^2 \right\rfloor \right)^+, \ i, j = \{1, 2\}.$$

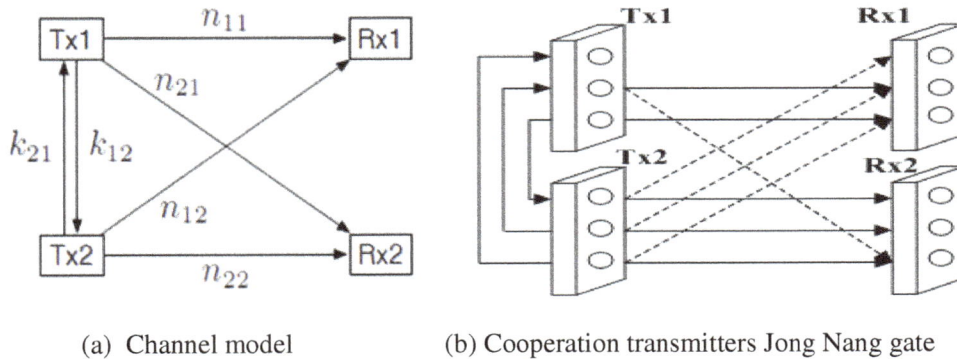

(a) Channel model   (b) Cooperation transmitters Jong Nang gate

Figure 9. Jong Nang gate IC with cooperation transmitters

Figure 9 shows Jong Nang gate with cooperation transmitters. A natural cooperative strategy between transmitters is that, prior to each block of transmission, two transmitters hold a conference to tell each other part of their messages [12]. Hence the messages are classified into two kinds: (1) cooperative messages, which are those known to both transmitters due to the conference, and (2) noncooperative ones, which are those unknown to the other transmitter since the cooperative link capacities are finite.

The generalization comes from making the mappings from X_1 to T_1 and from X_2 to T_2 in Figure 10.

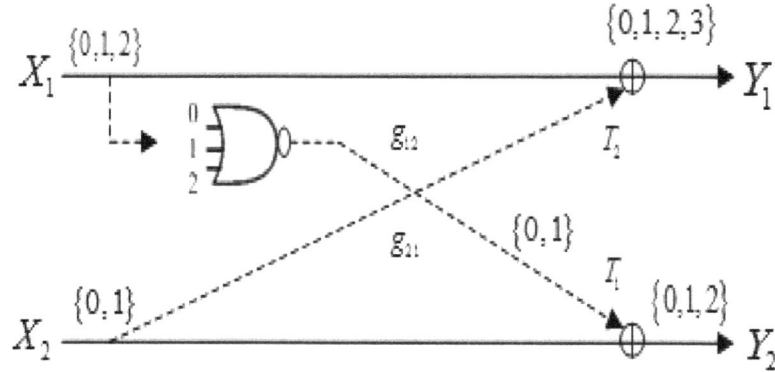

Figure 10. Block diagram of interference channel

The channel outputs are given by

$$Y_1 = y_1(X_1, T_2), \ Y_2 = y_2(X_2, T_1) \tag{32}$$

where $T_1 = t_1(X_1), T_2 = t_2(X_2)$, $X_1 = \{x_{10}, x_{11}, x_{12}\}$, and $X_2 = \{x_{20}, x_{21}, x_{22}\}$. We assume the functions y_1 and y_2 are injective in t_1 and t_2 respectively. We imply that $H(Y_1 | X_1) = H(T_2)$ and $H(Y_2 | X_2) = H(T_1)$. The probability of two transmitters are $P(T_1 | x_{10}, x_{11}, x_{12})$ and $P(T_2 | x_{20}, x_{21}, x_{22})$.

The Han-Kobayashi inner bound with the restriction that $p(u_1, u_2 | q, x_1, x_2) = p_{T_1|x_1}(u_1 | x_1)$ $p_{T_2|x_2}(u_2 | x_2)$ reduces to the following [13].

Proposition: Any rate pair (R_1, R_2) for the Jong Nang deterministic IC must satisfy the inequalities

$$R_1 \leq H(Y_1 | U_2, Q) - H(T_2 | U_2, Q), \tag{33}$$

$$R_2 \leq H(Y_2 | U_1, Q) - H(T_1 | U_1, Q), \tag{34}$$

We specialize the inner bounds to Gaussian as follows:

6.1 Time division

We obtain the time division inner bound on the capacity region of Gaussian IC such that

$$R_1 < \alpha C(SNR_1 / \alpha), R_2 < \overline{\alpha} C\left(SNR_2 / \overline{\alpha}\right) \ for \ \alpha \in [0,1] \tag{35}$$

6.2 Treating interference as noise

For treating interference as noise case, the capacity of region

$$R_1 < C\left(\frac{SNR_1}{1+INR_1}\right), \quad R_2 < C\left(\frac{SNR_2}{1+INR_2}\right) \tag{36}$$

6.3 Interference decoding

The capacity region of the Gaussian IC with strong interference is the set of rate pairs such that

$$R_1 \le C(SNR_1), \ R_2 \le C(SNR_2)$$

$$R_1 + R_2 \le \min\{C(SNR_1 + INR_1), C(SNR_2 + INR_2)\} \tag{37}$$

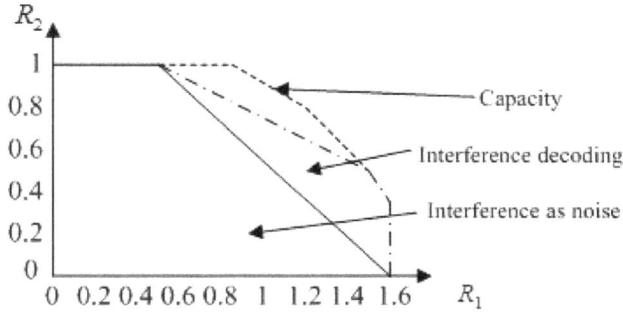

Figure 11. Capacity region and inner bounds

The above inner bounds are compared in Figure 11. It compares this inner bound to the capacity region given in [14] and to the region achievable by treating interference as noise. Interference decoding does not achieve the full capacity. Han-Kobayashi rate splitting and superposition coding are needed for full capacity. Finally, the combining functions and receiver functions are taken to be addition.

6.4 Feedback case

The scheme uses two stages. In the first stage, transmitter 1 and 2 send (a,b,c) and (d,e,f) respectively. Each receiver defers decoding to the second stage. In the second stage, using feedback, each transmitter decodes information of the other user: transmitter 1 and 2 decode (d,e,f) and (a,b,c), respectively. Each transmitter then sends information of the other user. Now each receiver can decode its own data from two received signals. Notice that the second stage was used for refining all bits sent previously, without sending additional information. In this case, we consider c and f. When we get output is one then feedback possible. Our logic is calculated in Table 4.

Table 4. Feedback case

Tx1 Input			Tx2 Input			Rx1 Output			Rx2 Output		
a	b	c	d	e	f	a	b	s	d	e	w
0	0	0	1	1	1	0	0	0	1	1	0
0	0	1	1	1	0	0	0	1	1	1	0
0	1	0	1	0	1	0	1	1	1	0	0
0	1	1	1	0	0	0	1	1	1	0	0
1	1	1	0	0	0	1	1	0	0	0	0

7. CONCLUSIONS

In this paper, we investigate the achievable symmetric rate for the Jong Nang gate interference channel. The deterministic models can be utilized to approximate the capacity of wireless communication networks and to design transmission strategies. The capacity of the JNG which means the wooden gate in Korea Jeju island dialect, had three wooden rafters placed on Jong-Ju-Mok (two large vertical stones with three holes) to convey the family's whereabouts was calculated. It is shown that this interference-decoding inner bound is tight under certain strong interference conditions. The inner bound is also shown to strictly contain the inner bound obtained by treating interference as noise, which includes interference alignment for deterministic channels. Practical JNG will be reasonable well matched to Tse [10] deterministic model of Gaussian IC.

ACKNOWLEDGEMENT

This work was supported by the World Class University (WCU) R32-2012-000-20014-0, BSRP 2010-0020942, MEST 2012-002521, NRF, Korea, the 2011 Korea-China International Cooperative Research Project (Grant No D00066, I00026) and [2]Hannam University while taking a sabbatical leave in SRM University, India.

REFERENCES

[1]   M. H. Lee, "Jong Nang (正木)", *EXPO '93 Information & Telecom*. Pavilion poster, 1993.

[2]   M. H. Lee, *The History of Information and Communication*, Kimyeong-Sa, Seoul, 1994.

[3]   M. H. Lee, "Jong Nang: The symbol of digital communication and Ying and Yang" *Telecom*, vol. 9, no. 1, 1993.

[4]   M. H. Lee, "Jong Nang System", Patent, no. 133285, Korea, 1998.

[5]   M. H. Lee, "The History of Jeju Jong Nang Binary Code", *IEEE VTS News*, vol. 50, no. 1, 2003.

[6]   M. H. Lee, X. Jiang, C. H. Choe, S. H. Kim, "Analysis of Jong Nang Multiple Access Channel', *ISITA 2006*, Seoul, Korea, 2006.

[7]   T. M. Cover, J. A. Thomas, *Elements of Information Theory*, John Wiley & Sons, New York, 1991.

[8]   C. E. Shannon, "A Mathematical Theory of Communication", *Bell System Technical Journal*, vol. 27, pp. 379-423 and 623-656, 1948.

[9]   N. Schrammar, "On Deterministic Models for Wireless Networks", Licentiate Thesis in Telecommunications Stockholm, Sweden 2011.

[10]   A. Salman, D. Suhas, D. Tse, "A Deterministic approaches to wireless relay networks", *ISIT*, 2007.

[11]   C. Suh, D. Tse, "Symmetric feedback capacity of the Gaussian interference channel to within one bit", *ISIT 2009*, Seoul, Korea, 2009.

[12]   I. H. Wang, D. Tse, "Interference mitigation through limited transmitter cooperation" , *IEEE Trans. on Information Theory*, vol. 57, no. 5, May 2011.

[13]   A. El Gamal, Y. H. Kim, *Network Information Theory*, Cambridge University Press, 2011.

[14]   B. Bandemer, A. El Gamal, "Interference decoding for deterministic channels", *IEEE Trans. Information Theory*, vol. 57, no. 5, May 2011.

6

THROUGHPUT ANALYSIS OF POWER CONTROL B-MAC PROTOCOL IN WSN

Ramchand V and D.K. Lobiyal

School of Computer and Systems Sciences,
Jawaharlal Nehru University, New Delhi, India
rchand.jnu@gmail.com
lobiyal@gmail.com

ABSTRACT

This paper presents a new methodology for energy consumption of nodes and throughput analysis has been performed through simulation for B-MAC protocol in Wireless Sensor Networks. The design includes transmission power control and multi-hop transmission of frames through adjusted transmitted power level. Proposed model reduces collision with contention level notification. The proposed model has been simulated using MATLAB. The simulations reveal better results for throughput of the proposed model as compared to B-MAC protocol. In this model we have included a mechanism for node discovery to find the location of the node before transmission of data to it. This increases the throughput of the network since the position of a dislocated node has been found, that results into successful transmission of frames. However, the energy consumption of a node increases due to energy consumed in node discovery.

KEYWORDS

Wireless Sensor Networks, Trigonometry, Possion distribution and Energy efficiency.

1. INTRODUCTION

Wireless communication is the most exciting area in the field of communication research. Over five decades it has been a major topic of research. Wireless Sensor Networks (WSNs) are an emerging technology that has become one of the fastest growing areas in the communication industry. They consist of sensor nodes that use low power consumption which are powered by small replaceable batteries that collect real world data, process it, and transmit the data to their destination nodes or a sink node or a server. WSN based applications usually have comfortable bandwidth requirement, the demand for using this medium is increasing with wide range of deployment for monitoring and surveillance systems as well as for military, Internet and scientific purposes. Wireless sensor networks will play an important role in future generation for multimedia applications such as video surveillance systems.

Transmission power control is provoked from potential benefits. The benefit is a more efficient use of the network resources. Allow a large number of simultaneous transmissions, power control increases the whole network capacity. Secondly energy saving is achieved by minimizing the average transmission power. The transmission power level is directly related to the power consumption of the wireless network interface. The lifetime of node's battery is becoming an important issue to the manufacturers and consumers, as devices are being used more frequently for transmission of signals\data packets\frames. It is becoming great interest to control the transmission power level of every node so that the lifetime of the wireless sensor network will be maximized.

Multiple access-based collision avoidance MAC protocols have made that a sender-receiver pair should first ensure exclusive access to the channel in the sender and receiver neighborhood before initiating a data packet transmission. Acquiring the floor allows the sender-receiver pair to avoid collisions due to hidden and exposed stations in shared channel wireless networks. The protocol mechanism used to achieve such collision avoidance typically involves preceding a data packet transmission with the exchange of a RTS/CTS (request- to-send/clear-to-send) control packet handshake between the sender and receiver. This handshake allows any station that either hears a control packet or senses a busy carrier to avoid a collision by deferring its own transmissions while the ongoing data transmission is in progress.

2. RELATED WORK

MAC layer has a vast impact on the energy consumption of sensor nodes. Communication is a major source of power consumption and the MAC layer design manages the transmission and reception of data over the wireless medium using the radio. The MAC layer is responsible for access to the shared medium. MAC protocols assist nodes in deciding when to access the channel. Pattern-MAC (P-MAC) for sensor networks adaptively determines the sleep-wake up schedules for a node based on its own traffic, and the traffic patterns of its neighbors. This protocol achieves a better throughput at high loads, and conserves more energy at light loads. In P-MAC, the sleep-wake up times of the sensor nodes is adaptively determined. The schedules are decided based on a node's own traffic and that of its neighbors. The improved performance of P-MAC suggests that 'pattern exchange' is a promising framework for improving the energy efficiency of the MAC protocols used in sensor networks [5].

S-MAC protocol achieves energy conserving through three basic techniques [10]. Nodes sleep periodically instead of listening continuously to an idle channel. Transceivers are turned off for the time the shared medium is used for transmission by other nodes overhearing is avoided, and a message passing scheme is used with the help of store-and-forward technique based on the buffer capacity. Each of the nodes has a fixed duty cycle. It can be used to tradeoff bandwidth and latency for energy conserving, but it does not allow adapting to network traffic. However, S-MAC protocol allows transmitting large messages by fragmenting them. It mitigates problems with higher delays and requires large storage buffers.

Power controlled Sensor MAC in wireless sensor network [9] provides a number of benefits. Sharing the medium efficiently therefore, decreasing the overall energy consumption of the network. Furthermore, increases the network capacity and maintains the network connectivity. A power controlled sensor MAC protocol addresses power controlled transmissions in wireless sensor network is an improvement to Sensor MAC protocol. It uses RTS (request to send), CTS (clear to send), DATA, and ACK (acknowledge), called as handshake mechanism. In order to save energy the frames are transmitted with the suitable power level, instead of the maximum power level.

3. PROPOSED METHODOLOGY OF B-MAC

3.1 Analytical model for power consumption

The function of network begins either because of node's mobility or node sensing an event. In order to find the location of dislocated node, "node discovery" will be performed. A node that senses an event, needs to calculate the contention level of nodes in the coverage area, estimate power required for transmission of control packets and frames, and probabilities of control packet and frames to be exchanged.

3.1.1 Discovering Nodes

The term "discovery" is used in many contexts to find the position of an object in physical space with respect to a specific frame of reference that varies across applications. Discovering nodes is the task of identifying two dimension or three dimension positions of the nodes.

Assumptions in discovering nodes:

I.   Sensor nodes are deployed statically equidistance to each other in the field to monitor events. Though nodes are mobile, we assume few nodes are stable within one hop communication range through which the location of other nodes can be estimated.

II. Nodes which are placed in the border of the field may go out of transmission or reception range is not given importance to locate their positions.

At the beginning of the network all the mobile nodes are placed at a specific location. All the nodes periodically broadcast a hello message to its neighbor nodes to gather its one-hop neighbor information such as node's location (varies over time) and power availability for communication.

A hello message contains the current list of its one-hop neighbors. Each sensor node has a certain area of coverage for which it can reliably and accurately report the particular quantity that it is observing. Due to mobility the density of network may vary over time. Rapid variation in the density of sensor node leads to frequent changes in the topology.

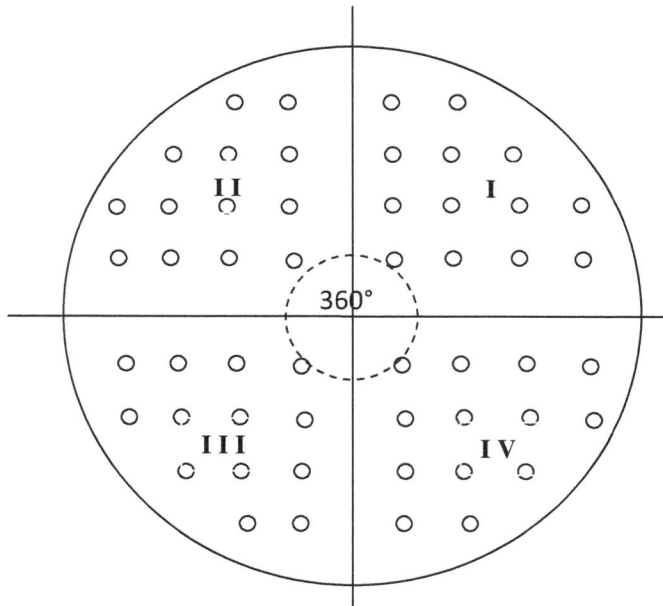

Figure 1. Network Scenario

The above figure is a sensor network were all the nodes are placed in a circular region with the center point of the circular region covering 360° which is divided in to four quadrants. Equal numbers of nodes are deployed equidistance to each other in all the quadrants.

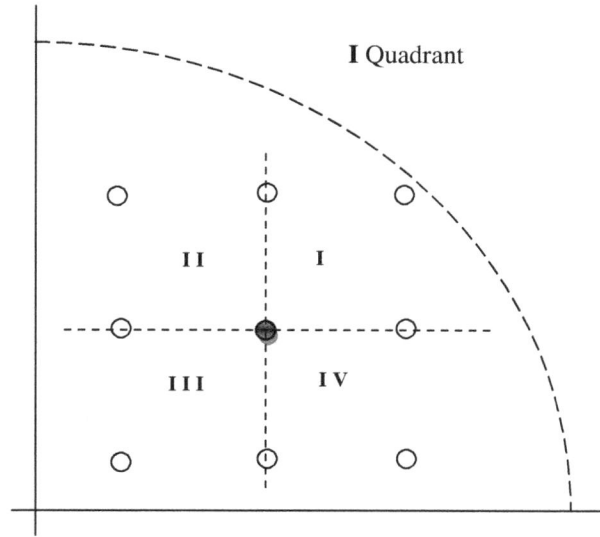

Figure 2. Nodes in I Quadrant

From figure 1 we took first quadrant nodes to find out the mobility of a node at time t with the assumption as stated above. As the network is divided in to four quadrants, every node in the network is surrounded by four quadrants with angles $\pi/2$, π, $3\pi/2$ and 0.

3.1.2 Locating angle of nodes

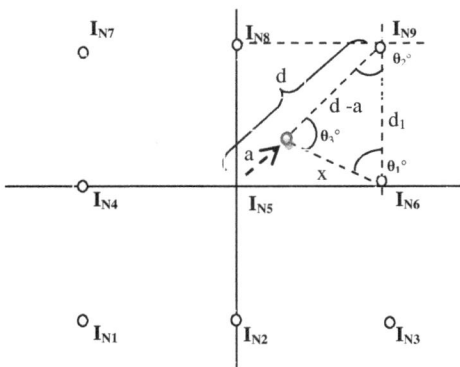

Figure 3a Node in I quadrant   Figure 3b Node in II quadrant

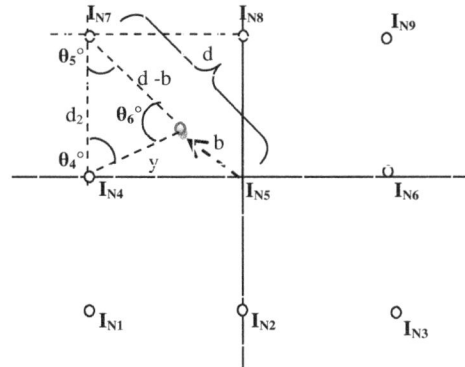

Figure 3a dislocation of a node is the first quadrant, where all the nodes are numbered as I_{N1}, $I_{N2}, \ldots, I_{N9}$. Node I_{N5} has moved towards its first quadrant. The location of node is calculated with the neighboring nodes I_{N6}, I_{N8} and I_{N9}.

$$\angle I_{N9} I_{N6} I_{N5} = 270° - \theta_2°$$

$$\angle I_{N6} I_{N9} I_{N5} = 180° - \theta_1°$$

$$\angle I_{N9} I_{N5} I_{N6} = ((270° - \theta_2°) + (180° - \theta_1°)) - 180° = \theta_3° \qquad (1)$$

d_1 is the distance between nodes I_{N6} and I_{N9}. Node I_{N5} has moved a distance from its original location. $d-a$ is the distance between nodes I_{N5} and I_{N9}. x is the distance between nodes I_{N5} and I_{N6}.

In figure 3b the same node I_{N5} has moved towards its second quadrant its new location is calculated with nodes I_{N4}, I_{N7} and I_{N8}.

$$\angle I_{N5} I_{N4} I_{N7} = 180° - \theta_4°$$

$$\angle I_{N5} I_{N7} I_{N4} = 270° - \theta_5°$$

$$\angle I_{N7} I_{N5} I_{N4} = ((270° - \theta_5°) + (180° - \theta_4°)) - 180° = \theta_6° \qquad (2)$$

d_2 is the distance between nodes I_{N4} and I_{N7}. Node I_{N5} has moved b distance from its original location. $d-b$ is the distance between nodes I_{N5} and I_{N7}. y is the distance between nodes I_{N5} and I_{N4}. Similarly same node may move to any quadrants and can be identified with trigonometry.

3.2 Transmission power control

The network function begins when a node senses an event and starts transmitting the sensed event in the form of message, data, frame or packet etc. The task of a node is to sense for events, transmit \ receive the data with other nodes, forward the data to a head node or sink node when ever required until the battery power drains. On a given time, either a node or few nodes may transmit out of N number of nodes deployed in the field.

Therefore, the probability of a node involved in transmission is $p_{n=1}(t) = 1/N$. Similarly the probability of more than one node involved in transmission is $p_{n=2,\ldots,N-1}(t) = N - 1/N$

The probability of transmitting nodes varies over time. Nodes active time, sleep time idle time as followed as per analytical model for power control T-MAC protocol [3]. At the initial stage of the network all the nodes are equipped with equal energy. Therefore, more number of nodes may employ themselves in sensing the events. Those nodes which have sensed some events will involved in forwarding of events as frames or signal, it leads to increase in high contention level among neighboring nodes. Any node before initiating a transmission estimates contention level to avoid collision with others.

$$C_L = A_N - \sum_{i=1}^{n} \pi r^2 * \frac{n_t}{N_A} \qquad (3)$$

C_L is the contention level
A_N is any node that measures the current contention level among neighbors.
$\sum_{i=1 \text{ to } n}$ are the nodes which are in contention to communicate.
πr^2 is the circular area where all the contending nodes reside.
n_t is the number of nodes contending at time t and

N_A is total number of nodes in a given area.

A node that wins the contention starts transmitting. Once the transmission is over the node goes to listen state. Further, nodes which are in the backoff mode wake up once the timers expire.

3.2.1 Contention Notification

Contention Notification (CN) messages alert the neighbor nodes not to act as hidden terminals when contention is high. Every node makes a local decision to send a CN message based on its local estimates of the contention level. Estimating contention level is either by receiving acknowledgment from the one-hop receiver or by measuring the carrier to noise plus interference ratio between the source and destination. Other way of estimating contention is by measuring the noise level of the channel. Any node in the network that has a frame to transmit senses the channel with the Clear Channel Assessment algorithm before initiating the transmission. When the noise level of the channel is higher than CCA threshold, the node takes random backoff. A node starts transmitting only when the noise level of the channel is smaller than CCA threshold. Noise level of a channel is measured by carrier to noise density ratio (CNDR),

$$CNDR = (E_f/n)*(N/C_A)*\frac{R}{B} \tag{4}$$

where E_f is the energy consumption in one frame transmission,
n is the noise level of current frame transmission,
N/C_A is number of nodes in given coverage area
R is the rate at which a frame is transmitted and
B is the channel bandwidth.

3.2.2 Power estimation of nodes

Source node transmit a frame to a destination node,

$$P_r = P_t\left(\frac{\lambda}{4\pi d}\right)^n \tag{5}$$

Given P_t is the transmit power of source and P_r is the power when the frame reaches the receiver. λ is the average arrival rate, and d is the distance from source to destination. n is the noise level of the channel. Source node sends RTS to destination node with the power level P_t, and the destination node receives the RTS package with the power level P_r then,

$$P_r = P_{frame}\left(\frac{\lambda}{4\pi d}\right)^n$$

$$\text{where} \quad P_{frame} = \frac{P_t R_s}{P_r}$$

where R_s is the sensitivity of received signal, sensitivity in a receiver is normally defined as signal s produce noise ratio at the transceiver / receiver node.

$$S_i = k(N_s + N_r)B * SNR$$

S_i is the Sensitivity
K is the Boltzman constant

N_s equivalent noise at source node
N_r equivalent noise at receiver node
B is bandwidth
SNR signal to noise ratio

Communication between source node to destination node with power level P_{frame} and the signal-to-noise radio (SNR) is not less than threshold Th, that utter the need for a minimal received power level,

$$\frac{P_{frame}}{N_{pr}} \geq Th$$

where N_{pr} is the noise power at receiver.

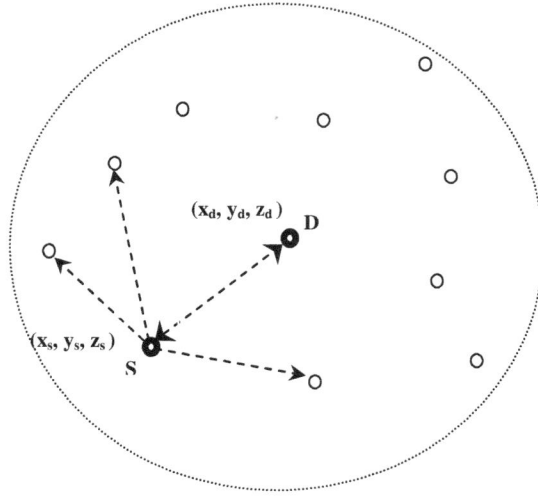

Figure 4. Dislocated nodes of WSN

Consider a source node at the location vector S = (x_s, y_s, z_s). The destination is located within the range of N+1 receiver at the location D = (x_d, y_d, z_d). The distance d from the source to one of the receivers in terms of the coordinates is

$$d_{sd} = \left| \vec{D} - \vec{S} \right| = \sqrt{(x_D - x_S)^2 + (y_D - y_S)^2 + (z_D - z_S)^2}$$

$$d_{sd} = \sqrt{(x_{D-S})^2 + (y_{D-S})^2 + (z_{D-S})^2} \tag{6}$$

Strength of the received frame is calculated with the use of log normal model of radio signal propagation between the source and destination nodes.

$$ss_D = TP_S - p_L - 10_n \log(d_{sd}) \tag{7}$$

ss_D is the received signal strength of the destination node.
TP_S is the transmitted power of source node.
P_L is Path loss, n is the path loss exponent.
d_{sd} is the distance between source and destination node, calculated from equation (6).

3.2.3 Adjusted transmission power

Let P_{max} is the max transmission power of nodes, and P is the current transmission power of node. E_{max} is the maximum energy level of nodes at the initial stage of the network operation begins, with $Pres$ is the residual power of a node. Optimal degree of a node is N, current degree of node is n.

To adjust the transmission power according to suitable degree, the adjusted transmission power P_{adj} is shown below,

$$P_{adj} = P + \left[\frac{N-n}{N} \right] * P$$

Further transmission power of node will be based on its residual power. Therefore, power available P_{AVAIL}

$$P_{AVAIL} = \left(\frac{P_{res}}{E_{max}} \right) * P_{max}$$

According to the residual power, improved adjusted power IP_{adj} is given as,

$$IP_{adj} = P_{adj} + \left[\left(\frac{P_{res}}{E_{max}} \right) * P_{max} \right] - P \tag{8}$$

3.2.4 Probabilities of control packets and frame exchange

When a node acquires frames for transmission at the rate λ_r and arrival follows Poisson distribution. On a frame to transmit it initiates the transmission with control packet exchange. To transmit an RTS packet a node takes μ_R time, therefore the average arrival rate of receiving an RTS packet $\lambda = \dfrac{1}{\mu_R}$. Therefore, probability of transmitting an RTS packet is calculated as

$$P(X=1) = \frac{e^{-(1/\mu_R)} * (1/\mu_R)^1}{1!} \tag{9}$$

On arrival of RTS packet the receiver calculate the power level of the transmitted control packet and reply the CTS packet with required power level. The probability of replying the CTS packet is represented as

$$P(X=1) = \frac{e^{-(1/\mu_C)} * (1/\mu_C)^1}{1!} \tag{10}$$

The node on receiving the reply it forwards the acquired frames to its one hop neighbor node. The frames are forwarded at the rate λ_r. Therefore, the power consumption of transmitting node is

$$P_t = p_r \lambda_r + \frac{e^{-(1/f)} * (1/f)^1}{1!} \tag{11}$$

where $\left(e^{-(1/f)} * (1/f)^1 \right)/1!$ is the probability of receiving a frame and p_r is the power consumed for receiving a frame.

On arrival of frame the receiver calculates the power level of the transmitted frame, it reply the ACK with required power level. The probability of replying the ACK is represented as

$$P(X=1) = \frac{e^{-(1/\mu_{ACK})} * \left(1/\mu_{ACK}\right)^1}{1!} \tag{12}$$

On arrival of ACK from the receiver the transmitting node sends the remaining frames with adjusted transmit power. Power consumption of a node transmitting frames with the adjusted power is given by

$$P_{t(IP_{adj})} = p_r \lambda_r + \frac{e^{-(1/f)} * (1/f)^n}{n!} \tag{13}$$

where $\left(e^{-(1/f)} * (1/f)^n\right)/n!$ is the probability of receiving n number of frames

4. SIMULATION RESULTS

The proposed model for estimating energy consumption in B-MAC protocol is implemented in mat lab. Proposed methodology has been designed with fewer numbers of mobiles nodes which are placed equidistance to each other. All the nodes are mobile by nature they relocated their positions subject to requirements of the tasks of their own or with the request of nodes with one hop communication range. The proposed protocol has been tested for unicast and broadcast communication with multi-hop transmission of frames. For unicast communication control packets like RTS-CTS are used. Adjusted transmission power level is used for broadcast communication. After broadcasting a frame, all nodes in the coverage area should refrain themselves from transmitting until one frame time has elapsed to allow transmitting the other node initiating a transmission is more efficient than control packet exchange for broadcast traffic. Frame length is varied as per the requirement of the application. The Simulation parameters used in the work are listed in the tables below

Table 1a Simulation parameter

Parameter	Value
Number of nodes	20
Contention window per slot duration	400 µs
Communication bandwidth	15 Kbps
Transmission Range	2 meters
Transmitting and Receiving antenna gain	Gt=1, Gr=1
Transmission power	0.031622777W
Carrier Sense Power	5.011872e- 12W
Received Power Threshold	5.82587e-09W
Traffic type	VBR
Initial Energy	500 Joule

Table 1b Frame parameters

Length (Bytes)	
Preamble	8
Synchronization	2
Header	5
Footer (CRC)	2
Frame length	Variable

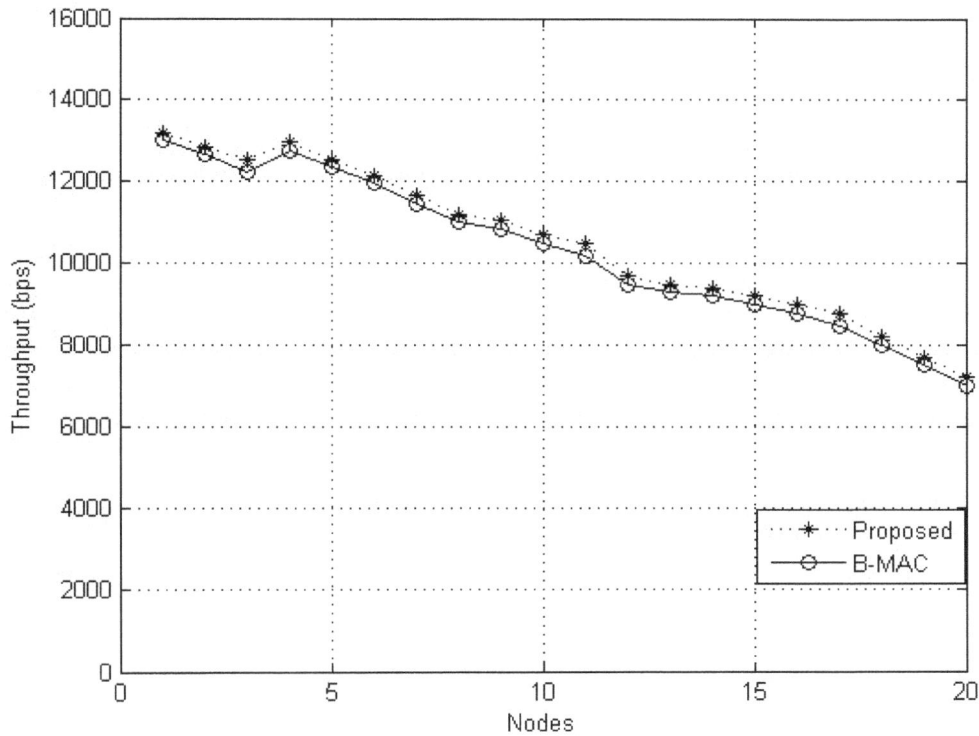

Figure 5 Throughput

The above figure shows the throughput of existing B-MAC protocol and the proposed model. Under low transmission rate unicast messages are exchanged with the use of control packet transmissions. Adjusted transmission power level is used for broadcasting frames. Proposed model deliver frames and achieves marginally better throughput for multi-hop transmission of frames than the existing work.

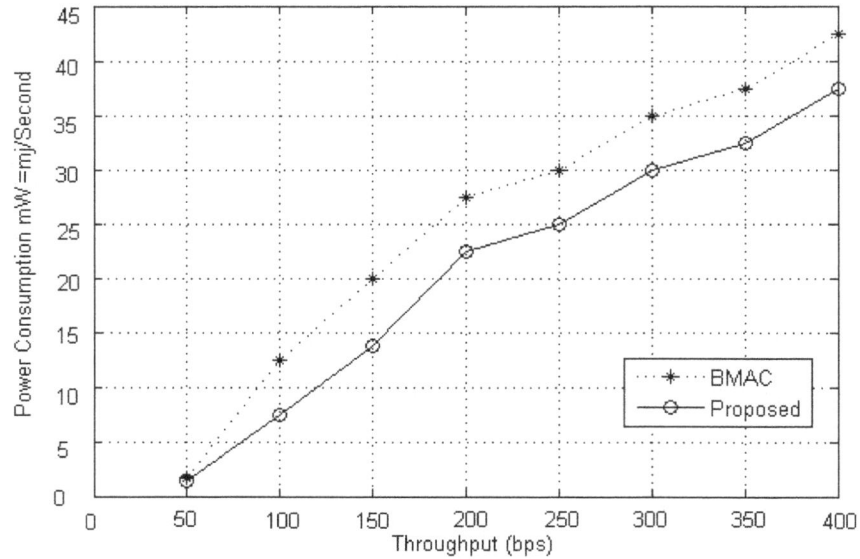

Figure 6 Power consumption

The above figure shows the comparison of energy consumption between B-MAC and proposed model. Energy efficiency is measured based on unicast, broadcast and multi-hop transmission of frames. While measuring the efficiency, sending rates are varied. The above figure presents the energy consumption of nodes involved in different duty cycles. As we observe in the multi-hop throughput, under low data rates, existing MAC has slightly lower throughput. The figure also shows the impact of the bits transmitted per second and power consumption of nodes in milli-watts. It is quite clear from the figure that the proposed work out performs the existing MAC in energy consumption.

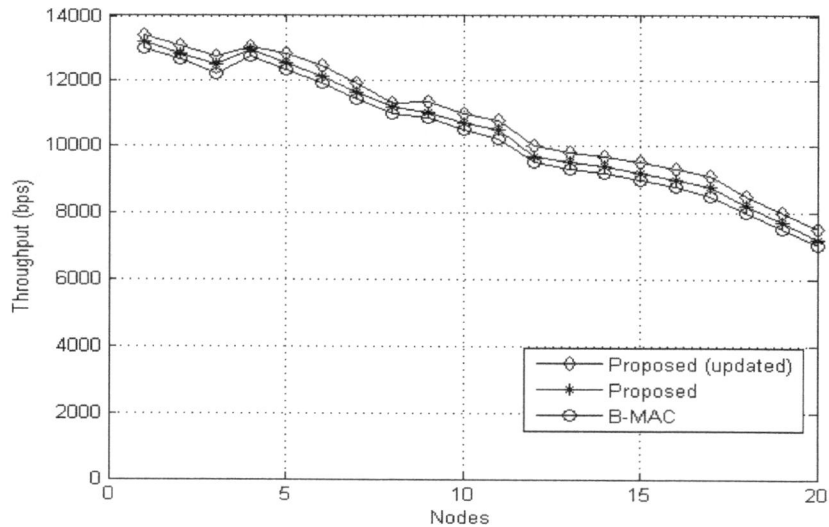

Figure 7 Improved Throughput

The above figure shows the throughput of existing B-MAC protocol, proposed model and proposed model with discovery of nodes. Proposed model deliver frames and achieves slightly better throughput than the existing work. Node discovery increases the overall throughput of the network.

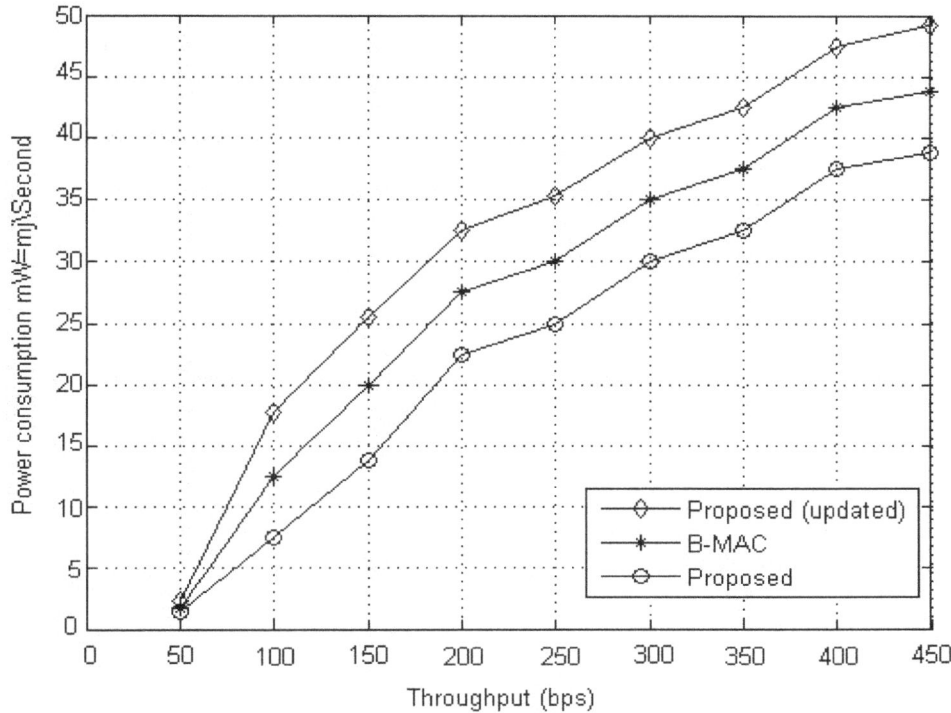

Figure 8 Power Consumption (depletion)

The above figure shows the comparison of energy consumption between B-MAC protocol, proposed model and the proposed model with node discovery. While discovering the dislocated nodes consume more energy. The figure shows the impact of energy depletion as compared with B-MAC protocol, power consumption of nodes is indicated in milli-watts.

5. CONCLUSION

In this work, we have proposed analytical model for estimating throughput of multi-hops and energy consumption in B-MAC protocol for Wireless Sensor Networks. The power consumption for an individual node is calculated for multi-hop communication. A node in the network saves its energy by changing its mode periodically. The proposed protocol shows better results than B-MAC protocol in terms of energy consumption. While designing the methodology for B-MAC, utilization variation in synchronization errors and transmission fairness and border nodes going away from the transmission are not focused, also when focusing on node discovery we obtaining better throughput but at the cost of power depletion of nodes. This may be explored in the future work.

REFERENCE

[1]   Andrea Goldsmith, Wireless Communication, Cambridge University press 2007.

[2]   C. Siva Ram Murthy, B.S.Manoj, Ad hoc Wireless Sensor Networks.

[3]   Ramchand V and D.K.Lobiyal, "An Analytic model for Power Control T-MAC protocol", International Journal of Computer Applications (0975-8887) Volume 12-No.1, December 2010.

[4]   Yaser Mahmood A. Hamid, and D. K. Lobiyal "IPCM/COMPOW: An Efficient Power Saving Scheme for Multi-Hop Wireless Ad Hoc Networks", ICWN 2008, pp.452-458.

[5]   Youngmin Kim, Hyojeong Shin and Hojung Cha, "Y-MAC An Energy Efficient Multi-channel MAC Protocol for Dense Wireless Sensor Networks", International Conference of Information Processing in Sensor Networks 2008.

[6]   I. Rhee, A. Warrier, M. Aia, J. Min and M.L. Sichitiu, "Z-MAC: a hybrid MAC for wireless sensor networks," in IEEE/ACM Transactions on Networking, vol. 16, no. 3, pp. 511-524, 2008.

[7]   V. Rajendran, K. Obraczka and J.J. Garcia-Luna-Aceves, "Energy-efficient, collision-free medium access control for wireless sensor networks," in Wireless Networks, vol. 12, pp. 63-78, 2006.

[8]   P. Ji, C. Wu, Y. Zhang and Z. Jia, "Research of an energy-aware MAC protocol in wireless sensor network," in Chinese Control and Decision Conference, pp. 4686-4690, 2008.

[9]   P.C. Nar and E. Cayirci, "PCSMAC: "A power controlled sensor MAC protocol for wireless sensor networks", Wireless Sensor Networks". Proceedings of the Second European Workshop on 31 Jan.-2 Feb. 2005 Page(s):81 – 92.

[10] Wei Ye, John Heidehann, Deborah Estrin, "An Energy Efficient MAC protocol for Wireless Sensor Networks", INFOCOM 2002, IEEE Computer and Communication Socities Proceedings, Volume 3

7

PERFORMANCE EVALUATION OF TCP VARIANTS IN WI-FI NETWORK USING CROSS LAYER DESIGN PROTOCOL AND EXPLICIT CONGESTION NOTIFICATION

Manish Devendra Chawhan[1] and Dr Avichal R.Kapur[2]

[1]Assistant Prof, Dept of E&C, SRKNEC, India
mchawhan76@yahoo.com
[2] Dean(QA) &Advisor, NYSS,MGI, India
avichal.kapur@yahoo.co.in

ABSTRACT

TCP was mainly developed considering assumption of wired network, ignoring the properties of wireless transmission. Wireless transmission links are highly unreliable causing loss of packets all the time. The proper approach to dealing with lost packets is to send them again, and as quickly as possible. This paper aims at studying the effects of unidirectional and bidirectional networks on various TCP variants. The effect of application of SNOOP and ECN on the performance enhancement of TCP along with TCP variants is assessed, improving the performance of TCP over wireless network by implementing cross layer design protocol (Snoop). ECN is used to avoid congestion and Snoop aims at retransmitting the lost packets from base station, avoiding retransmission from the transmitter. The performance of different TCP variants such as TCP Tahoe, Vegas, Reno, New Reno, Sack are analyzed on Wi-Fi scenario. These results can be analysed from throughput and congestion window plots in the paper. The simulator used for implementation in Network Simulator-2 (NS2).

KEYWORDS

Explicit Congestion Notification (ECN) Transmission Control Protocol (TCP), Snoop Protocol, Network Simulator-2(NS)

1. INTRODUCTION

In recent years, issues regarding the behavior of TCP in high-speed and long-distance networks have been extensively addressed. The packet loss in heterogeneous network environment into three categories:
(1) packet loss due to overflow in intermediate routers.
(2) packet loss due to high bit-error-rate in wireless links.
(3) packet loss due to user mobility (e.g. handoff).
The well known problem of TCP in high bandwidth delay product networks is that the TCP Additive Increase probing mechanism is too slow in adapting the sending rate to the available bandwidth. Various TCP Variants[1,2] have been suggested for this, such as TCP Vegas, Reno, NewReno, Tahoe, Sack. The performances of various variants are analyzed in bidirectional scenarios using throughput and congestion window plots. For the purpose of analyzing the effects of reverse traffic, i.e. congestion and other losses due to wireless environment, a scenario of bidirectional wifi network has been created and all the simulation results have been tested on the same scenario.

The first kind of packet loss has been taken into consideration in traditional TCP design and implementation. The congestion control mechanisms in Reno TCP are aimed to tackle this kind of packet loss. When the wireless links become parts of TCP connection, the second and third kind of packet loss occur, which break the assumption that packet loss is only caused by the congestion in the intermediate routers and thus it could degrade the TCP performance. Snoop protocols [3, 4] can significantly improve the TCP performance in that it hides the second kind of packet loss from the TCP sender by means of local retransmission and local timeout mechanism at base station. Moreover, Snoop protocol addresses the third kind of packet loss by using routing technology. Snoop protocol [5] is to alleviate degradation in performance of TCP over heterogeneous network. They improve the end-to-end performance on networks with wireless links without changing existing TCP implementations at hosts in the fixed network and without recompiling or re-linking existing applications. They achieve this by a simple set of modifications to the network-layer (IP) software at the base station. These modifications consist mainly of caching packets and performing local retransmissions across the wireless link by monitoring the acknowledgments to TCP packets generated by the receiver. Snoop protocol can achieve speedups of up to 20 times over regular TCP in the presence of bit errors on the wireless link. It also is significantly more robust at dealing with multiple packet losses in a single window as compared to regular TCP. Snoop protocol is a cross layer design protocol i.e. transports aware link layer protocol [6].

1.1 Related Work

In paper [2] the author have incorporate non congestion-related random losses and round-trip delay in this model, and show that one can generalize observations regarding TCP-type congestion avoidance to more general window flow control schemes. They consider explicit congestion notification (ECN) as an alternate mechanism (instead of losses) for signalling congestion and show that ECN marking levels can be designed to nearly eliminate losses in the network by choosing the marking level independently for each node in the network. While the ECN marking level at each node may depend on the number of flows through the node, the appropriate marking level can be estimated using only aggregate flow measurements, i.e., per-flow measurements are not required.

The throughput of an user using ECN marks is much better (about 5 times) than a user without ECN marks. This improvement in performance is due to the user attributing all losses to random losses in the network. Since, the marking level makes sure that there are very few congestion related losses, most of the packet losses seen by the user are indeed due to random losses.

In [7], have described the design and implementation of a protocol, called the snoop protocol, which improves TCP performance in wireless networks. The protocol modifies network-layer software mainly at a base station and preserves end-to-end TCP semantics. The main idea of the protocol is to cache packets at the base station and perform local retransmissions across the wireless link. The experiments show that it is significantly more robust at dealing with unreliable wireless links as compared to normal TCP. The throughput speedups achieved of up to 20 times over regular TCP in experiments with the protocol.

In [5] paper includes a simulation-based performance analysis of the most important TCP versions over wireless networks. In addition, analyzing those TCP versions in the same environment but including the Snoop protocol. In the paper the segments sequence numbers vs. time graph for all TCP versions considered over the plain wireless network. From the graph it is noticed that all TCP versions perform as expected. However, it is found that in TCP Vegas, not many analyses are available to conclude about its behaviour. They were in the process of identifying the ground causes for this behaviour analyzing the congestion window and how the error model affects Vegas. Initial analysis tells us that Vegas is affected more than the other TCP versions whenever the length of the error bursts is more than four packets.

TCP Vegas, is the best performing version with Snoop, and trying to find the reasons behind this behaviour and the behaviour of other TCP Variants. They had stated that, there are unknown interactions between the snoop protocol and these TCP versions that need a more detailed investigation.

The above work showed that ECN improves TCP performance in Congestion related losses and Snoop improves in wireless related losses. Hence in [8],[9],[10]analysis were done to improve the performance of TCP using both ECN and Snoop both applied in a Wi-Fi and Wi-Max Scenario which showed improved result.

Also, the performance and behaviour of all TCP Variants with the Snoop protocol needed to be analyzed and hence this paper investigates the behaviour of TCP Variants with and without the Snoop Protocol. ECN is also applied to Vegas (E-Vegas) to further improve its performance in a unidirectional network.

2. WI-FI NETWORK SCENARIO

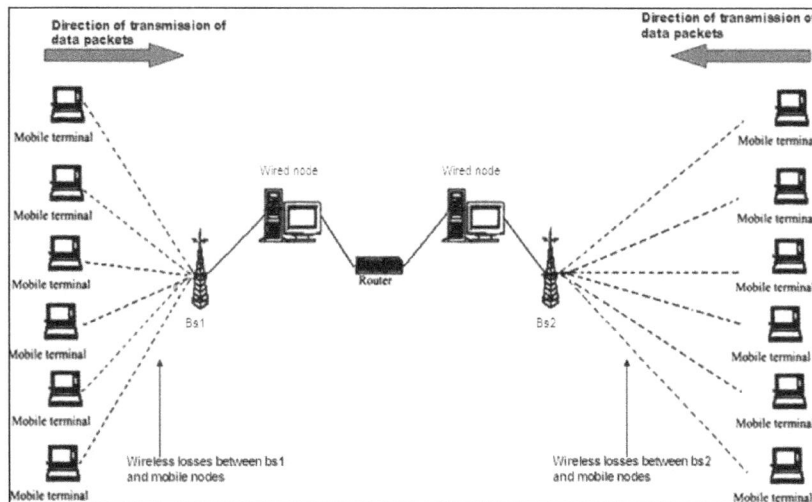

Fig 1. WiFi Network Scsenario

The Scenario shown in Fig 1 is a WiFi unidirectional/bidirectional network .It consists of 12 wireless nodes. There are 2 wired cum wireless base stations which are BS1 and BS2. The LAN bandwidth is of 10 Mbps. LAN may be implemented with Snooping agent. LAN nodes are connected to base station with a 1Mbps 1msec RED bidirectional link. Red link is for marking the packets in case congestion occurs [11]. Wireless nodes of base station 1 (BS1) are sending data to wireless nodes of base station 2 (BS2) and vice versa. The arrows in the figure shows the direction in which node transmits data .The protocols used are TCP variants such as Tahoe, Reno, New Reno, Vegas & Sack. They are attached with a File Transfer Protocol (FTP). The wireless routing protocol used is DSDV [12]. All the nodes start transmission simultaneously leading to congestion in base station 1 (BS1) and base station 2 (BS2). Wireless losses are introduced between base station 2 (BS2) and wireless nodes connected to base station 2 (BS2). Also wireless losses are present between base station 1 (BS1) and nodes connected to base station 1 (BS1). Here there are wireless losses introduced at both the ends and there is a very high possibility of congestion occurring at the base stations because of the high transmission rates. These two problems are then overcome using ECN and SNOOP.[8,9]The analysis is done for all TCP variants in unidirectional scenario where only nodes connected to base station 1

(BS1) are transmitting to nodes connected to base station 2 (BS2) and then the same analysis is done for the WiFi bidirectional scenario as shown in figure 1.

3. SIMULATION RESULTS OF WI-FI NETWORK:

3.1 Network parameters

Parameters	Value	Meaning
Channel	Wireless channel	channel type
Adhoc Routing	DSDV	Routing protocol
Error data packet size	100kb	
Wireless Error rate	5 %	
LAN bandwidth	10 Mbps	
Ifq	Drop tail	Type of queuing used
Application	FTP	
Mac standard	802.11	
Ifqlen	50	max packet in ifq
Ant	Omni Antenna	antenna model
opt(x)	600	X dimension of the topography
opt(y)	600	Y dimension of the topography
opt(netif)	Phy/WirelessPhy	network interface type

Table 1. Network Parameters

3.2 Simulation results for Wi-Fi unidirectional network

We will now analyse simulation results for various TCP Variants along with 'EVegas' (ECN [11] with Vegas) for the scenario as described above, however in this part we do not implement 'Snoop' on the given network.

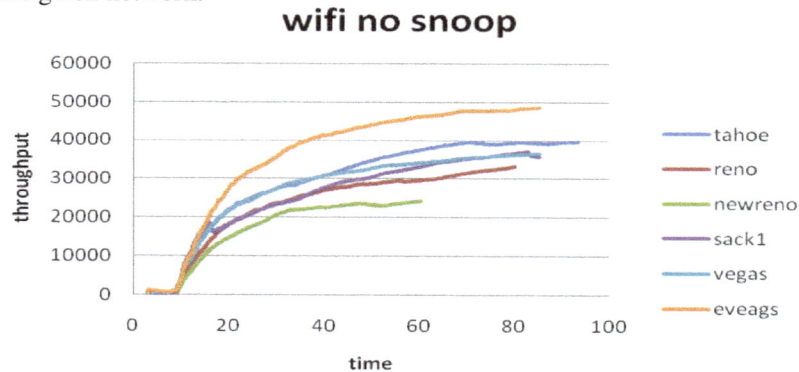

Figure 2. Throughput vs Time plot for TCP Variants in Wifi Unidirectional network (Without Snoop).

Above graph is the throughput plot for the given scenario with WIFI(IEEE 802.11). In the above figure various TCP variants such as NewReno, Reno, Sack1, Tahoe, Vegas, 'EVegas' are used. This is a plot for throughput vs. time which is total number of packets delivered per unit time. Throughput is calculated in bps(bits per second). The result was worst for TCP New Reno which is shown in green colour. The throughput for 'EVegas' (TCP variant is Vegas along with the congestion control protocol ECN)[11] is the best of all. E-Vegas reach to a maximum of (approx.) 49kbps which is much better than New Reno, Tahoe or 'Vegas'. Percentage increase of throughput of E-Vegas with respect to Vegas is 40%. However it was found out that the performance can be further improved using SNOOP on the above network.

The following figure shows the simulation result for a network on which SNOOP was implemented

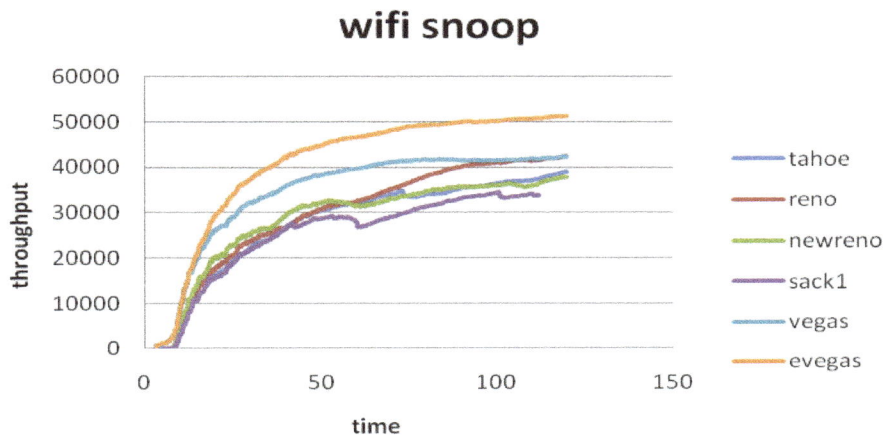

Figure 3. Throughput vs Time plot for TCP Variants in Wifi Unidirectional network (With Snoop).

In figure 3 various TCP variants such as, NewReno, Reno, Sack1, Tahoe, Vegas, EVegas [1]. Graphs are plotted for throughput Vs time. It can be observed that with the application of SNOOP over the network, the throughput of all the variants increases significantly. The However throughput for E-VEGAS reaches to a maximum of (approx.) 51kbps that is much better than other TCP variants, as shown in the graph. The maximum throughput achieved in case of 'Evegas', without SNOOP is 49Kbps and with SNOOP is 51Kbps. Percentage increase of throughput of E-Vegas with respect to Vegas is 4%. Hence it can now be concluded that 'EVegas' works better in both types of networks.

3.3. Analysis of Results

The same result can also be justified by using congestion window plot for all the TCP variants. Following figure shows the congestion window plot for the different variants. (Only the best four TCP variants have been considered.)

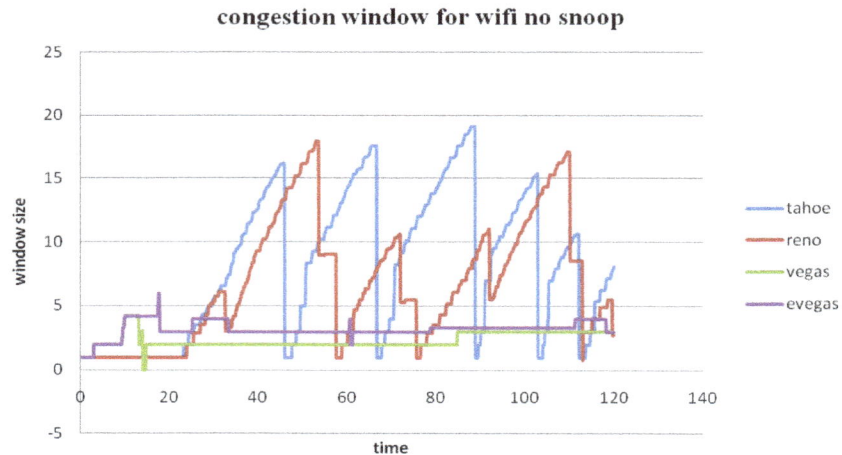

Figure 4. Congestion window plot for TCP variants in Wifi Unidirectional network.(Without Snoop)

The above graph (figure 4) is the congestion window plot for Wi-Fi unidirectional network without SNOOP. It is found that, though the window sizes of different TCP's are greater than the window size of TCP Vegas or 'EVegas', still the throughput of TCP Vegas was better than most of the variants and performance of 'EVegas' is the best. This is because the congestion window resets for a greater number of times in other TCP variants as compared to TCP Vegas and E-Vegas. In these cases it maintains low and steady transmission rate. However the better performance of 'EVegas' over TCP Vegas can be attributed to the fact that the size of window for 'EVegas' is greater than that of TCP Vegas.

Similarly, the congestion window plots for the same four TCP variants were plotted for a network on which SNOOP was implemented.

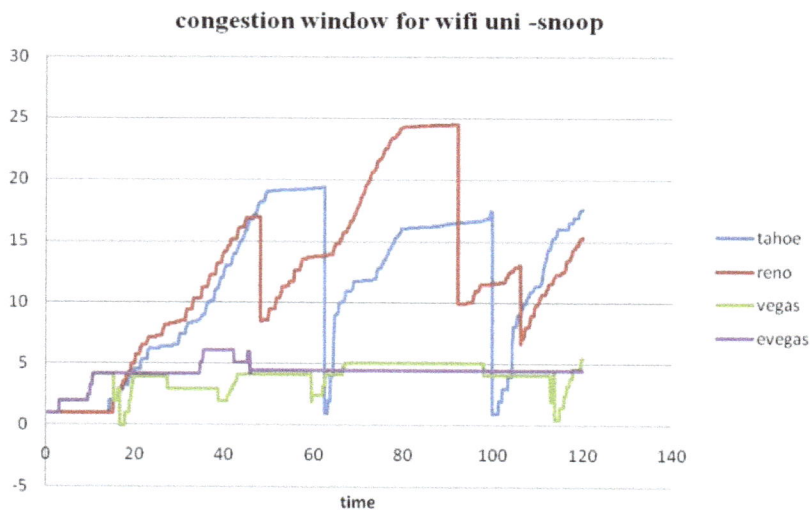

Figure 5. Congestion window plot for TCP variants in Wifi Unidirectional network.(With Snoop)

Now analyzing figure 5, that the number of window resets in case of all TCP variants have reduced drastically. This accounts for the increased rate of throughput as seen from the diagram above (figure 3.). This is primarily due to the implementation of SNOOP, which doesn't all the backward transmission of duplicate acknowledgement to the source and hence the source does not have to reduce its window size as well as transmission speed even when the packet is lost at the receiver side.

It can also be seen that in case of TCP-Vegas or 'EVegas' it maintains a low but constant transmission rate. On observing the two figures (4,5) it can be seen that the window size of TCP Vegas and 'EVegas' in figure (6) is higher. Hence the performance of TCP-Vegas and 'EVegas' is better than all the other TCP Variants. However in TCP Vegas we can see that the window size goes to half its actual size because of congestion losses. As we already know that these congestion losses can be taken care of applying ECN[11] over Vegas. Hence in the graph for ECN Vegas (EVegas) we see that it does not reset at all. Which is the main reason why the performance of EVegas is better than Vegas. The RED queuing mechanism helps to detect congestion before it occurs and ECN notifies the sender for it.

4. SIMULATION RESULTS FOR BIDIRECTIONAL WIFI NETWORK

The analysis of simulation results for various TCP Variants along with 'E-Vegas' and 'E-New Reno' for the bidirectional scenario is described below. ('E-Vegas' and 'ENewReno' are TCP Variants 'Vegas' and 'NewReno' implemented with ECN.)

Throughput vs. Time Plot for TCP Variants in a wifi bidirectional scenario (without Snoop):

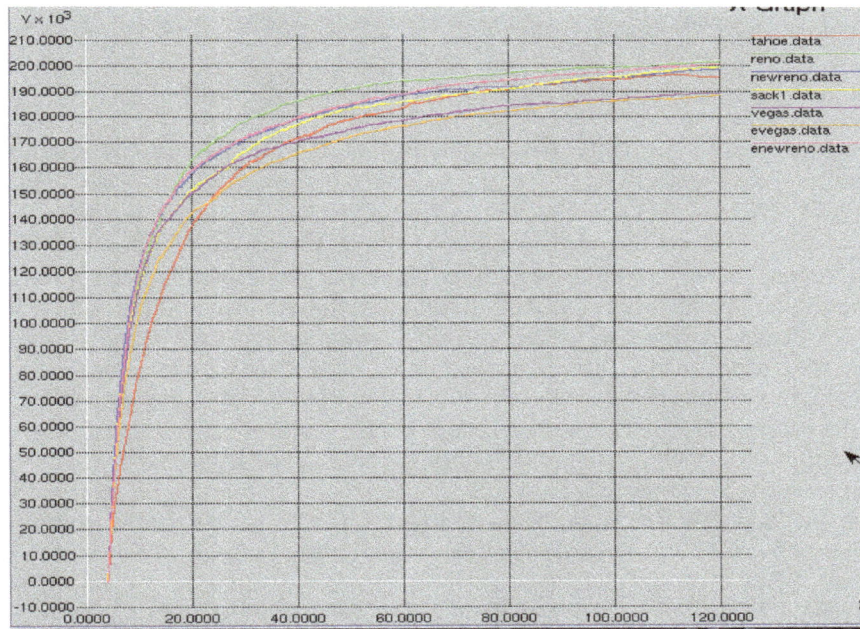

Fig 6. Throughput Vs Time for TCP Varients without Snoop

Throughput vs. Time Plot for TCP Variants in a wifi bidirectional scenario (with Snoop):

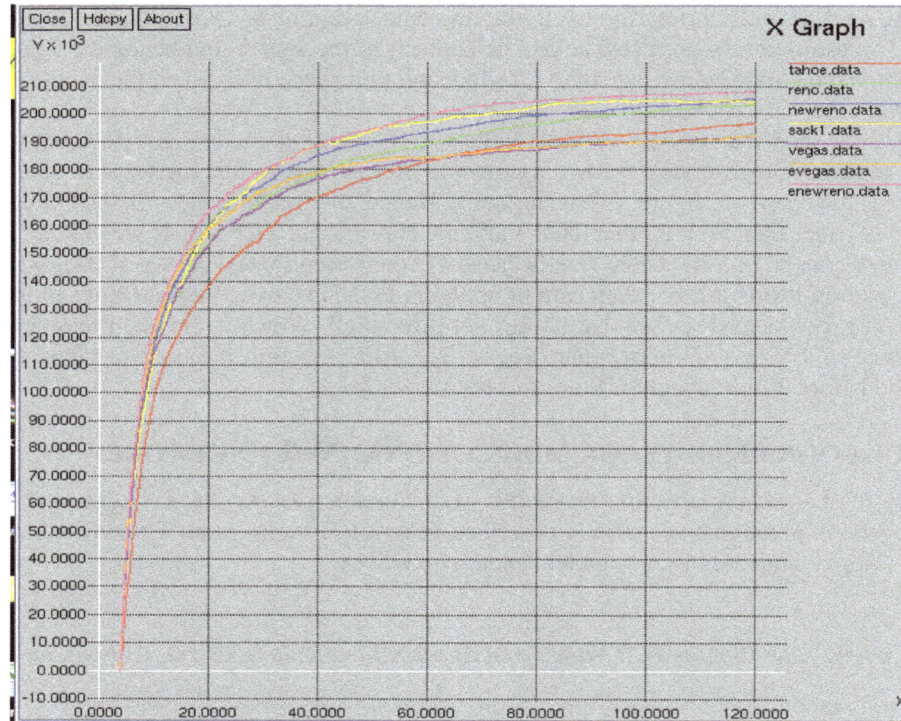

Fig 7. Throughput Vs Time for TCP Variants with Snoop

Fig 6 is the throughput plot for the given scenario (Fig 1) Here various TCP variants[13] are used. According to the analysis done on unidirectional networks it was seen that the performance of 'EVegas' was the best amongst all TCP variants, hence the testing of all the TCP variants is done in bidirectional scenario by implementing ECN with them. However in this case it is seen that the performance of New Reno was better than the other variants (figure 6), this is because of the modifications in TCP Reno which were incorporated in New Reno.[14] So to further enhance its performance a congestion control protocol (ECN)[8] is applied to this network and the performance of New Reno along with ECN (E-NewReno) is analysed.

Figure 7 is the throughput plot for all TCP Variants with cross layer design protocol SNOOP applied to the network(fig 1). The performance of all the variants has improved; this is because of the application of SNOOP protocol. It is observed that the performance of E-New Reno can be regarded as the best amongst all the variants. The maximum throughput achieved in case of 'ENewReno', without SNOOP is 200Kbps and with SNOOP is 208Kbps It is been found that there is approximately 4% improvement in the performance of E-NewReno, after the application of SNOOP protocol to the network. Hence we observe that for bi directional scenario the best TCP variant is TCP NewReno and E-NewReno is even better.

4.1. Analysis of Results using Congestion Window Plots

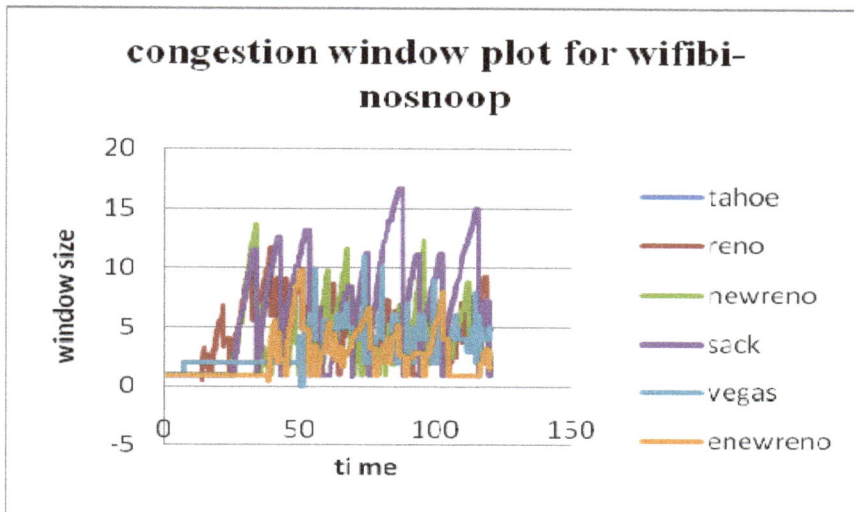

Fig 8. CWND Vs Time without Snoop

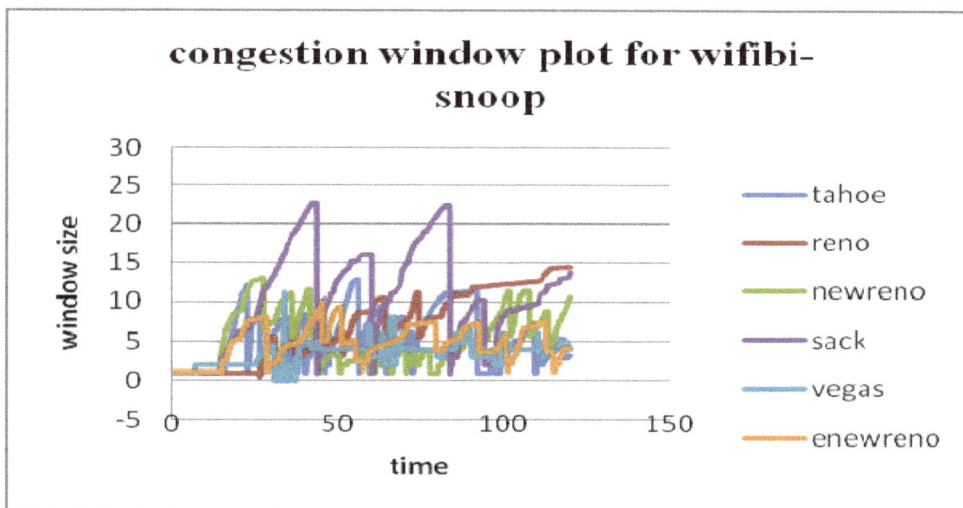

Fig 9. CWND Vs Time with Snoop

Fig 8 & 9 are congestion window plots for bi-directional network of various TCP variants. This graph helps us to further evaluate, why the performance of NewReno has improved whereas the performance of Vegas has degraded.

Vegas connection on the forward path additively shrinks the cwnd thus obtaining a poor utilization of the bottleneck link. Hence to optimize the full bandwidth Vegas needs to send packets at a fast rate, however due to its congestion control mechanism it is not able to transmit packets at a greater rate but it keeps a slow and steady flow of packets. Therefore the performance of Vegas degrades in a bidirectional networks [16].

In case of Bidirectional network, there is heavy traffic flow in both the directions. There also occurs loss of packets in wireless environments which is best overcome using a protocol which rapidly re-transmits the lost packets. However if congestion occurs at any intermediate bottle neck node then fast retransmission may lead to more congestion. Hence this congestion problem is best avoided by using ECN (Explicit Congestion Notification) along with TCP Variants. NewReno is seen to have a good transmission rate and it also works well when ECN is applied with it. This is also visible from the graphs where it can be seen that though the window size of Newreno is not the greatest, but it avoids window resetting when ECN is applied to it. This is the reason why the performance of 'E-NewReno' is seen to be better. Secondly the throughput performance of all the variants also increases with the application of cross layer protocol Snoop.

5. CONCLUSION

The analysis of the result shows improvement in throughput of Vegas and E-Vegas with and without snoop with respect to other TCP Variants ie Reno, Newreno, Sack, and Tahoe.

The analysis of the result for unidirectional network shows improvement in throughput of E-Vegas (49kbps) by 40% with respect to Vegas (35 kbps) (Fig 2.) and further the performance of TCP Vegas is improved by application of SNOOP in the Wi-Fi scenarios. The Throughput of all Variants increases and the throughput of E-Vegas reaches 51kbps(fig 3) which shows further 4% increase in the performance of TCP Vegas when snoop is applied.. ECN helps in congestion control and SNOOP will retransmit the packets that are lost from nodes in between, saving nearly half the retransmission time and avoiding the decreasing in transmission speed and an optimum transmission performance in a wireless network can be achieved. TCP Vegas is better than most of the TCP Variants and 'EVegas' is the best combination of variants for a unidirectional network, with as well as without SNOOP.

In bidirectional scenario the reverse traffic significantly affects the behaviour of protocols. Here from the research results that are achieved, it could be said that the performance of TCP NewReno was better than the other TCP variants. The throughput of Newreno without snoop is 180kbps(Fig 6.) and throughput with snoop reaches to 205kbps (fig 7) which shows improvement by 13.8% . This shows that Newreno shows much more improvement than all other TCP Variants. Also the application of ECN with NewReno (E-Newreno) further increases throughput performance in a bidirectional network.(Fig 7.)

6. FUTURE WORK

Further research is being carried out on the same direction on Wi-Max Networks.

7. REFERENCES

[1] A Comparative Analysis of TCP Tahoe, Reno, New-Reno, SACK and Vegas L.S. Brakmo & L.L. Peterson IEEE Journal on Selected Areas in Communication, vol. 13[1995],(1465- 1490).

[2] "End-to-End Congestion Control Schemes: Utility Functions, Random Losses and ECN Marks" Srisankar Kunniyur and R. Srikant Coordinated Science Laboratory University of Illinois kunniyur@uiuc.edu; rsrikant@uiuc.edu. Research supported by NSF Grants ANI-9813710 and ANI-9714685.

[3] Performance Evaluation of TCP over WLAN 802.11 with the Snoop Performance Enhancing Proxy Chi Ho Ng, Jack Chow, and Ljiljana Trajkovic School of Engineering Science Simon Fraser University Vancouver, British Columbia Canada.

[4] TCP performance in case of bi-directional packet lossR.E. Kooij and R.D. van der Mei TNO Information and Communication Technology, P.O. Box 5050, 2600 GB Delft, The Netherlands This

work is partially funded by the Dutch Ministry of Economic Affairs under the programme TS ICT Doorbraakprojecten', project TSIT 2031 EQUANET.

[5] Performance of TCP over Wireless Networks with the Snoop Protocol . Sarma Vangala and Miguel A. Labrador Department of Computer Science and Engineering University of South Florida, Tampa, U.S.A fvangala,labradorg@csee.usf.edu Proceedings of the 27th Annual IEEE Conference on Local Computer Networks (LCN.02) 0742-1303/02 © 2002 IEEE.

[6] Cross-layer Design for Wireless Networks. Sanjay Shakkottai, Theodore S. Rappaport and Peter C. Karlsson□ Wireless Networking and Communications Group (WNCG), Department of Electrical and Computer Engineering, The University of Texas at Austin.

[7] "Improving TCP/IP Performance over Wireless Network", Hari Balakrishnan, Srinivasan Seshan, Elan Amir and Randy H. Katz (hari,ss,elan,randy}@CS.Berkeley.EDU Computer Science Division University of California at Berkeley.

[8] A paper titled "Performance Enhancement of TCP using ECN and Snoop Protocol for Wireless Network" Manish D Chawhan and Dr. Avichal Kapur was published as a research paper in "The 2009 International Conference on Wireless Networks(ICWN-2009) organized by WORLDCOMP'09(world academy of science) held on July 13-16,2009 in Las Vegas Nevada, USA.

[9] "TCP Performance Enhancement using ECN & Snoop Protocol for Wi-Fi Network", Manish D Chawhan and Dr. Avichal Kapur.The 2nd IEEE International Conference on Computer and Network Technology (ICCNT'10), 2010, Bangkok, Thailand.

[10] "Performance Enhancement of TCP using ECN and Snoop Protocol for Wi-Max Network",Manish D Chawhan and Dr. Avichal Kapur published in "The   International   Journal of Computer Applications",Volume 1, Number 16 - Article 5, ISSN No 0975-8887.

[11] Extended ECN: A New Mechanism for Improving TCP Performance Over Wireless Links *George T. Plataniotis and Angelos N. Rouskas* Department of Information and Communication Systems Engineering, University of the Aegean, Karlovassi 83200, Samos, Greece.

[12] Destination-Sequenced Distance Vector (DSDV) Protocol. Guoyou He Networking Laboratory Helsinki University of Technology.

[13] TCP Variants and Network Parameters: A Comprehensive Performance Analysis. Md. Shohidul Islam, M.A Kashem, W.H Sadid, M. A Rahman, M.N Islam, S. Anam. Proceedings of the International MultiConference of Engineers and Computer Scientists 2009 Vol I IMECS 2009, March 18 - 20, 2009, Hong Kong ISBN: 978-988-17012-2-0

[14] The NewReno Modification to TCP's Fast Recovery Algorithm, Network Working, Group S. Floyd and T. Henderson, ACIRI. U.C. Berkeley.

[15] A SIMULATION OF ECN-CAPABLE MULTICAST MULTIMEDIA DELIVERY IN NS-2 ENVIRONMENT Robert R. Chodorek Department of Telecommunications The AGH University of Technologyal. Mickiewicza 3030-059 Kraków Poland chodorek@kt.agh.edu.pl Proceedings 14th European Simulation Symposium A. Verbraeck, W. Krug, eds. (c) SCS Europe BVBA, 2002

[16]Performance Evaluation and Comparison of Westwood+, New Reno, and Vegas TCP Congestion Control.Luigi A. Grieco and Saverio Mascolo *Dipartimento di Elettrotecnica ed Elettronica, Politecnico di Bari, Italy* a.grieco,mascolo@poliba.it

[17] Approaches of Wireless TCP Enhancement and A New Proposal Based on Congestion Coherence .Chunlei Liu ,Lucent Technologies ,6100 E Broad St, Columbus & Raj Jain ,Department of Computer and Information Science ,Ohio State University, Columbus. Proceedings of the 36th Hawaii International Conference on System Sciences (HICSS'03)0-7695-1874-5/03 © 2002

[18] End-to-End Performance Evaluation of Selected. TCP Variants across a Hybrid Wireless Network. A.O. Oluwatope, A. B. Obabire, and G. A. Aderounm at Comnet Laboratory, Obafemi Awolowo University. Ile-Ife, Nigeria

[19] TCP Performance Enhancement Over Wireless Network. Aiyathurai Jayananthan, Doctor of Philosophy University of Canterbury New Zealand 2007.

[20] NS-2 simulation tool home page.

[21] "Performance Evaluation and Comparison of Westwood+, New Reno, and Vegas TCP Congestion Control" Luigi A. Grieco and Saverio Mascolo *Dipartimento di Elettrotecnica ed Elettronica, Politecnico di Bari, Italy* a.grieco,mascolo@poliba.it

[22]"Approaches of Wireless TCP Enhancement and A New Proposal Based on Congestion Coherence" Chunlei Liu ,Lucent Technologies ,6100 E Broad St, Columbus & Raj Jain ,Department of Computer and Information Science ,Ohio State University, Columbus. Proceedings of the 36th Hawaii International Conference on System Sciences (HICSS'03)0-7695-1874-5/03 © 2002

[23]"Improving TCP Performance over WiMAX Networks Using Cross-layer Design" , Jin Hwang, Sang Woo Son, Byung Ho Rhee , Department of Computer & Communication Engineering, Hanyang University. 17 Haengdang-dong, Seougdong-gu, Seoul, 133-791, Korea. Third 2008 International Conference on Convergence and Hybrid Information Technology.

[24]"Performance Enhancement Techniques for TCP, Over Wireless Links", E. Yanmaz, S.-C. Wei, and O. K. Tonguz, Department of Electrical and Computer Engineering, Carnegie Mellon University, Pittsburgh, PA 15213-3890, USA.

[25]."Cross Layer Communication for Wireless Networks", Satish Ket, Vijay Shinde, Ravindra Khandare, R.N.Awale. International Conference on Advances in Computing, Communication and Control (ICAC3'09).

[26] "A cross-layer design for TCP end-to-end performance improvement in multi-hop wireless networks," Rung-Shiang Cheng, , Hui-Tang Lin. 2008 Elsevier.

[27] " Cross-layer congestion control in ad hoc wireless networks," Dzmitry Kliazovich, Fabrizio Granelli, Department of Information and Communication Technologies, University of Trento, Via Sommarive 14, I-38050 Trento, Italy. 2005 Elsevier.

[28] "A Cross-layer Scheme for TCP Performance Improvement in Wireless LANs" ,Dzmitry Kliazovich and Fabrizio Granelli ,DIT - University of Trento ,Via Sommarive 14, I-38050,Trento, ITALY . Globecom 2004 IEEE Communications Society.

[29] "TCP Performance Enhancement in Wireless/Mobile Communications", S. Hadjiefthymiades, S. Papayiannis, L. Merakos, Communication Networks Laboratory, Department of Informatics and Telecommunications, University of Athens, Athens 15784, Greece.

[30]" Cross-Layer Design Tutorial" Frank Aune - faune@stud.ntnu.no Norwegian University of Science and Technology, Dept. of Electronics and Telecommunications, Trondheim, Norway. Published under Creative Commons License.

8

Performance Simulation and Analysis for LTESystem Using Human Behavior Queue Model

Tony Tsang

Hong Kong Polytechnic University
Hung Hom, Hong Kong.
Email: ttsang@ieee.org

Abstract

Understanding the nature of traffic has been a key concern of the researchers particularly over the last two decades and it has been noticed through extensive high quality studies that traffic found in different kinds of IP/wireless IP networks is human operators . Despite the recent findings of real time human behavior in measured traffic from data networks, much of the current understanding of IP traffic modeling is still based on simplistic probability distributed traffic. Unlike most existing studies that are primarily based on simplistic probabilistic model and traditional scheduling algorithms, this research presents an analytical performance model for real time human behavior queue systems with intelligent task management traffic input scheduled by a novel and promising scheduling mechanism for 4G-LTE system. Our proposed model is substantiated on human behavior queuing system that considers real time of traffic exhibiting homogeneous tasks characteristics. We analyze the model on the basis of newly proposed scheduling scheme for 4G-LTE system. We present closed form expressions of expected response times for real time traffic classes. We develop a discrete event simulator to understand the behavior of real time of arriving tasks traffic under this newly proposed scheduling mechanism for 4G-LTE system . The results indicate that our proposed scheduling algorithm provides preferential treatment to real-time applications such as voice and video but not to that extent that data applications are starving for bandwidth and outperforms all other scheduling schemes that are available in the market.

1. INTRODUCTION

In the Internet, Quality of Service (QoS) management allows different types of traffic to contend inequitably fornetwork resources. Bandwidth is the key heuristic to manage real life network utilities like video and voice overremote locations. Three main QoS frameworks such as IntServ, DiffServ and MPLS have been introduced to providedifferential treatment to a variety of applications available in real time service internet [1] . The differentiation ofmultiple classes of traffic is fundamentally relied on these frameworks that utilize various queuing and schedulingcombinations for separating different traffic classes. Further, the traffic separation is categorized under specificparameters like bandwidth, delay, jitter and packet-loss rate. The different arrangements of these parameters can bebundled under variety of queuing and scheduling methods. It is therefore vital to QoS frameworks that modelingof traffic behavior through network domains is accurate so that resources can be optimally assigned.

Understanding the nature of traffic has been a key concern of the researchers particularly over the last twodecades and it has been noticed through extensive high quality studies that traffic found in different kinds ofIP/wireless IP networks is human operators [2] . Despite the recent findings of real time human behavior inmeasured traffic from data networks, much of the current understanding of IP traffic modeling is still based onsimplistic Poisson distributed traffic. In this paper, we add to a more realistic modeling of network domains throughthe following main contributions: (1) the presentation of an analytical approach and closed form expressions tomodel the accurate behavior of multiple classes of wireless IP traffic based on a human behavior queuing systemunder real time assumptions, (2) the derivation of expected waiting times of corresponding human behavior trafficclasses and formulation of an embedded intelligent task management and (3) the detailed simulation results to giveexact QoS parameter bounds to validate the analytical framework.

We have analyzed the traditional scheduling schemes based on real time human behavior queuing system, whereasin current study, we analyze the newly proposed scheduling scheme to guarantee tight bound QoS to all kind of trafficin human behavior service wireless internet. The rest of the paper is structured as follows. Section II summarizesrelated work. The Real Time Human Behavior Queue Model have been discussed in Section III. The simulationand analysis results are given in Section IV. Finally, Section V concludes this work.

2. RELATED WORK

Queuing theory is the backbone of telecommunication systems. The major concern about internet traffic is: howburstiness (commonly known as human behavior operations) behavior can be managed in real time spans. Theexperimental queuing analysis and simulation studies with human behavior packet data arrival traffic have beenperformed in [3] and [4] respectively. These studies merely indicate that providing hard and tight bound guaranteesfor different QoS parameters such as maximum delay, delay-jitter and cell-loss probabilities in the presence ofhuman behavior traffic is nontrivial especially if the coefficient of variation of the marginal distribution is large.The readers are referred to [5, 6] to get a detailed overview of other queuing based results available in the presenceof human behavior traffic. The core limitation of these findings is based on the fact, that FIFO logic has beenconsidered to understand the behavior of traffic, which can't be used to provide differential treatment to multipleclasses of traffic with different QoS time constraints.

The desire to dispense divergent QoS guarantees to different classes of customers in wireless Internet is leadingto the use of priorities in tenns of allocation of resources. Multiple priority based classes are supported by the IProuters and ATM switches. The authors in [7] have used human-in-the-loop (analytical) Model to provide numericalresults for two different classes of traffic input based on Real Time Task Release Control Process (RT-TRCP). Anotable discrepancy of RT-TRCP is the estimation of large set of parameters. It has been shown that Real Time TaskRelease Control Process (RT-TRCP) can prioritize each class in its own buffer [8]. The flow control managementbased on the computation of probability of various types of traffic classes has been discussed in [9]. The otherwork related to this study can be found in studies [10, 11] . Unfortunately, in related work, the issue of providingQoS guarantees to the end-user based on tight bound QoS parameters has not been properly addressed.

In addition, we refer the readers to [12, 13] regarding the work that has been carried out in terms of IP networkperformance evaluation. The analysis conducted in [12, 13] has two main disadvantages; first the reported queuingmodels did not employ human-in-the-loop phenomenon for network traffic input and second, they have only usedsingle class of traffic for conducting analysis by neglecting the performance affect of other subsequent traffic classes.To overcome the limitations of related work, they presented a novel analytical framework [14, 15] based on realtime human behavior queuing system, that contemplates task management of human behavior traffic. In the relatedwork, they analyzed the traditional scheduling schemes such as priority and round robin. It is well known thattraditional scheduling schemes can't provide the required QoS to all types of traffic found in modem wirelessnetworks. Hence, in this current study, we analyze real time human behavior model on the basis of a novel andmost promising scheduling mechanism titled as, "Best Scheduling Algorithm (BSA)" and find exact packet delaysfor the corresponding classes of human behavior traffic. The results indicate that BSA completely outperforms alltraditional and other available scheduling schemes. To date, no closed form expressions have been presented forreal time human behavior model with such scheduling mechanism.

3. REAL TIME HUMAN BEHAVIOR QUEUE MODEL

The basic elements of Real Time Human Behavior Queue Model are its actions, which represent activities carriedout by the systems being modeled, and its operators, which are used to real time descriptions.

Time point

A time point is a time instant with respect to the global clock of the system; it does not have duration. It specifiesthe starting and stopping times of an action. Using a time point, we can instruct the system to generate an actionat a particular point in time. Time point progresses consistently in all parts of the system. More formally, the timepoint is defined by using a discrete time domain, which contains the following properties:

$$\forall t \exists t' t < t' \land \forall t'' : t < t'' \Rightarrow t' \leq t''$$

We assume a fixed set of clocks $t = \{t_0, \dots, t_i\}$. The special time point t_0, which is called the start time point, always has the value 0.

Time Constraint

An action can exist for a period of time; this duration is called the time constraint of the action. A time constrainthas a starting and an ending point. It consists of a lower-bound and an upper-bound time point, where the lowerboundtime point enables an action in a module, and the upper-bound time point disables the action at that pointin time. Formally, we define a time constraint in the following:

$$\mathcal{T}_i = \{[\tau_{i_{min}}, \tau_{i_{max}}] \mid \forall t_i \in T\} \text{with } 0 \leq \tau_{i_{min}} \leq \tau_{i_{max}}.$$

Timed Action

A timed action is a tuple $< \alpha, \lambda, \mathcal{T} >$consisting of the type of the action α, the rate of the action λ and temporalconstraint of the action $\mathcal{T}$. The type denotes the kind of action, such as transmission of data packets, while therate indicates the speed at which the action occurs from the view of an external observer. The rates are used todenote the random variables specifying the duration of the actions. The actions can be defined in different types ofprobability distribution function such as human behaviors distribution. Moreover, each transition is also boundedby a temporal constraint.

Real Time Single Server Queue Model

Consider the following single-server queue model. Tasks arrive periodically, at rate λ, i.e., a new task arrives every $1/\lambda$ time units. The tasks are identical and independent of each other and each task brings w units of work, where w is an independent identically distributed (i.i.d.) random variable whose probability distribution is f_W with bounded support $[W_1, W_2]$ for some $W_1 > 0$ and $W_1 \leq W_2$. In the rest of the paper, we will assume this bounded support assumption on f_W without explicitly repeating it. Let w be the mean of w with respect to f_W. Let δ_w be the Dirac delta distribution centered at $\overline{w}$. We will use the δ distribution for the scenario when the tasks are homogeneous. Note that the task arrival process under consideration is deterministic. We briefly discuss the implications of stochastic inter arrival times in Section. The tasks must be serviced in the order of their arrival. We next state the dynamical model for the server, which specifies the state-dependent service times for the server.

3.1   Real Time Server Model

Let $x(t) \in [0,1]$ be the server state at time t, and let $b: \mathcal{R} \to 0,1$ be such that $b(t)$ is 1 if the server is busy at time t, and 0 otherwise, where $\mathcal{R}$ is the set of real numbers. The evolution of $x(t)$ is governed by a simple first-order model

$$x(t) = \frac{b(t) - x(t)}{\tau}, x(0) = x_0 \tag{1}$$

where $\tau > 0$ is a time constant that determines the extent to which past utilization affects the current state of the server, and $x_0 \in [0,1]$ is the initial condition. The quantity $x(t)$ bounded by time interval $[t_1, t_2]$ denotes the utilization ratio of the server, i.e., the fraction of the recent history when the server was busy. Physically, $x(t)$ represents the perceived workload of the operator based on its recent utilization history within the time interval $[t_1, t_2]$. Equation (1) can be considered to be the continuum limit of the discrete time exponential moving window average by rewriting the time derivative in (1) from first principles

$$x(t + \Delta t) \approx \left(1 - \frac{\Delta t}{\tau}\right) x(t) + \frac{\Delta t}{\tau} b(t). \tag{2}$$

A simple moving window average model has been proposed in [16] for computing the utilization ratio. For other models of human mental workload, we refer the reader to [17]. The time constant τ corresponds to the inverse of the sensitivity of the operator to its recent utilization history: larger τ correspond to lower sensitivity and smaller τ correspond to higher sensitivity. Note that the set $[0,1]$ is invariant under the dynamics in (1) for any $\tau > 0$ and any $b: \mathcal{R} \to \{0,1\}$.

The service times bounded by time interval $[t_1, t_2]$ are related to the state $x(t)$ through a map $\mathcal{S}: [0,1] \to \mathcal{R}_{>0}$, where $\mathcal{R}_{>0}$ is the set of positive real numbers. If a task is allocated to the server at state x, then the amount of time required to perform unit work is given by $\mathcal{S}(x)$. Therefore, if the amount of work associated with a task allocated to the server at state x is w, then the service time on that task is $w\mathcal{S}(x)$. This linear decomposition of the total service time within time interval $[t_1, t_2]$ into the amount of work associated with the task and the rate of performing work with respect to the initial server state is an approximation to a more realistic scenario where the rate of performing work also depends on the amount of work such a model has been proposed in [4]. The linear decomposition that we use is reasonable especially for small heterogeneity in the tasks. In our framework, the controller cannot interfere with the server while it is servicing a task. Hence, the only way in which the server state can be controlled is by scheduling the beginning of service of tasks after their arrival. Such controllers are called task-release controllers and will be formally characterized later on. In this paper, we assume that $S(x)$ is positive valued, continuous, and convex. Let $S_{min} := min\{S(x)|x \in [0,1]\}$, and $S_{max} := max\{S(0), S(1)\}$

The solution to (1) is $x(t) = e^{-t/\tau}(\oint_0 (1/\tau)b(s)e^{s/\tau}ds + x_0)$. This implies that the server state $x(t)$ is increasing when the server is busy, i.e., when $b(t) = 1$, and decreasing when the server is not busy, i.e., when $b(t) = 0$. Note that $S(x)$ is not necessarily monotonically increasing in x, since it has been noted in the human factors literature [18] that, for certain cognitive tasks demanding persistence, the performance [which in our case would correspond to the inverse of $S(x)$] could increase with the state x when x is small. This is mainly because a certain minimum level of human behavior mental arousal is required for good performance. A well-known empirical law capturing such characteristics is the Yerkes-Dodson law [18] . A loose experimental justification of this server model in the context of human-in-the-loop systems is included in the related work [19], where $S(x)$ for that setup was found to have a U-shaped profile. We will use that particular $S(x)$ from [19] for various numerical illustrations in this paper. We provide further experimental evidence for this model in Section . It is important to note that the U-shaped relationship between the service time and the operator's utilization, as would be dictated, for example, by the Yerkes-Dodson law, falls within our assumptions on $S(x)$ within time interval $[t_1, t_2]$ but it is not essential. In particular, our assumptions on $S(x)$ also allow it to be monotonically increasing, decreasing, or even constant over $x \in [0,1]$.

3.2   Real Time Task Release Control Policy

We now describe task release control policies within the time interval for the Real Time Human Behavior queue. Without explicitly specifying its domain, a task release controller u acts like an ON-OFF switch at the entrance of the queue, e.g., see Fig. 1. In short, u is a task release control policy if $U(t) \in \{ON, OFF\}$ for all $t > 0$, and an outstanding task is assigned to the server if and only if the server is idle, i.e., when it is not servicing a task and $u = ON$. Let $\mathcal{U}$ be the set of all such task release control policies. For a given $\tau > 0$ and f_W , let $n_u(t, \tau, \lambda, f_W, x_0, n_0)$ be the queue length, i.e., the number of outstanding tasks, at time t, under task release control policy $u \in \mathcal{U}$, when the task arrival rate is λ and the server state and the queue length at time $t = 0$ are x_0 and n_0 , respectively. For brevity in notation, we will sometimes use the short hand notation $n_u(t)$ to denote the queue length at time t under task release control policy u when the other parameters are clear from the context. Note that we allow $\mathcal{U}$ to be quite general in the sense that it includes control policies that are functions of $\lambda, S, x(t), f_W, \tau, n_u$, bounded by the time interval $[t_1, t_2]$ etc.

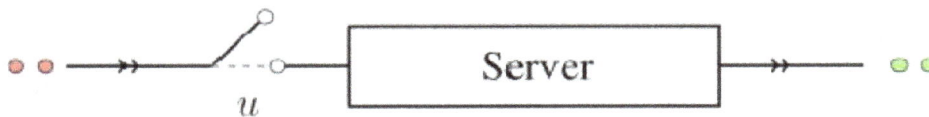

Figure 1: Real -Time Task Release Control Architecture

3.3   Define Human Behavior Stabilizable Arrival Rate

We now formally state the problem. Define the maximum stabilizable arrival rate within time interval $[t_1, +\infty]$ for policy u as

$$\lambda_{max}(\tau, F_W, u) :=$$

$$= sup\{\lambda : \lim_{t \to +\infty} supn_u(t, \tau, \lambda, f_W, x_0, n_0)$$

$$< +\infty \forall_{x_0} \in [0,1], n_0 \in \mathcal{N} \, a.s.$$

The quantity $\lambda_{max}(\tau, f_W, u)$ will also be referred to as the throughput under policy u within time interval $[t_1, t_2]$. The maximum stabilizable arrival rate over all policies, or simply the real time throughput, is defined as $\lambda_{max}^*(\tau, f_W) := sup_{u \in \mathcal{U}} \lambda_{max}(\tau, f_W, u)$. For a given $\tau > 0$ and f_W , a task release control policy u is called maximally stabilizing if, for any $x_0 \in [0,1], n_0 \in \mathcal{N}$, $\lim sup_{t \to +\infty} n_u((t, \tau, \lambda, f_W, x_0, n_0) < +\infty$ for all $\lambda \le \lambda_{max}^*(\tau, f_W)$ within time interval $[t_1, +\infty]$ almost surely. The primary objective in this paper is to compute the real time throughput and design a corresponding maximally stabilizing task release control policy for the dynamical queue whose server state evolves according to (1), and where $S(x)$ is positive, continuous, and convex.

In this paper, we extensively focus on a specific class of task release control policies threshold policies. For a given $x^* \in [0,1]$, the x^* -threshold policy is defined as

$$u_{x^*}(t) = \begin{cases} ON, \\ if x(t) \le x^* \\ OFF, \\ otherwise. \end{cases}$$

We prove that an appropriate threshold policy is maximally stabilizing when the tasks are homogeneous and utilize the threshold policies in the time interval to prove bounds on the real time throughput when the tasks are heterogeneous.

3.4   Simple Bounds on the Real Time Throughout

We start by deriving simple bounds on the real time throughput.

Proposition II.1: For any $\tau > 0$ and f_W , we have that $\lambda_{max}^*(\tau, f_W) \in [(\overline{w}S(1))^{-1}, \overline{w}S_{min})^{-1}]$

Proof: The time between the start of service of successive tasks consists of two parts: the time to actively service a task, and the time when the server is idle, as governed by the task release control policy. The upper bound on the throughput is obtained by neglecting the idle times and by assuming that the server spends the least amount of time to service every task. The lower bound is proven by considering the trivial policy $u(t) \equiv ON$ as follows. Assume, by contradiction, that the queue length grows unbounded under this policy for some initial condition for an arrival rate $(\overline{w}S(1))^{-1} - \varepsilon$ for some $\varepsilon > 0$. For a queue length growing unbounded in the time interval $[t_1, +\infty]$, the server state exceeds $1 - \eta$ for any given $\eta > 0$ in some finite time $T[t_1, t_2]$. Note that the queue length remains bounded until $T[t_1, t_2]$. After $T[t_1, t_2]$, all the service times per unit work are upper bounded by $S(1) + \theta$ where $\theta \ge 0$ depends on η through the continuity of $S(x)$. One can select η and hence θ such that

$$(\overline{w}S(1) + \overline{w}\theta)^{-1} > (\overline{w}S(1))^{-1} - \varepsilon. \tag{3}$$

By the strong law of large numbers, with probability one, the average service time per task after T is upper bounded by $\overline{w}S(1) + \overline{w}\theta$. Combining this with (3), we get that, after T, the arrival rate is strictly less than the mean service time with probability one and hence the queue length cannot grow unbounded with the time constraint $[t_1, t_2]$. This contradiction proves that the queue length remains bounded with probability one for an arrival rate $(\overline{w}S(1))^{-1} - \varepsilon$ for any ε and for any initial condition, which in turn proves that $\lambda^*_{max}(\tau, f_W)$ is lower bounded by $(\overline{w}S(1))^{-1}$.

The bounds with the time constraint $[t_1, t_2]$ obtained in Proposition II.1 can be shown to be tight for some simple cases. Consider first the case when $S \equiv \varepsilon$ for some constant $\varepsilon > 0$. In this case, $S(1) = S_{min} = \varepsilon$ and hence Proposition II.1 implies that $\lambda^*_{max}(\tau, f_W) = (\overline{w}\varepsilon)^{-1}$ for all $\tau > 0$. Additionally, the trivial policy $u(t) \equiv ON$ is maximally stabilizing. Another simple case is when $S(x)$ is nonincreasing. In this case, $S(1) = S_{min}$ and hence Proposition II.1 implies that $\lambda^*_{max}(\tau, f_W) = \overline{w}S(1))^{-1}$ for all $\tau > 0$. One can show that the trivial policy $u(t) \equiv ON$ on is maximally stabilizing in this case as well. We now derive tighter bounds on the real time throughput and design corresponding maximally stabilizing task release control policies with time constraint.

3.5   Real Time Arriving Tasks

In this subsection, we consider the special case when the arriving tasks are homogeneous with time constraint, i.e., every task brings in exactly the same deterministic amount of work with it. Formally, we let $f_W(w) = \delta_{\overline{w}}(w)$ for some $\overline{w} \in [\mathcal{W}_1, \mathcal{W}_2]$. We start by studying specific types of equilibria that are associated with the trivial policy $u(t) \equiv ON$.

1)One Task Equilibria: Let x_i be the server state with the time t_i at the beginning of service of the i th task and let the queue length be zero at that instant. The server state upon the arrival of the $(i = 1)$ task is then obtained by integration of (1) over the time period $[0, 1/\lambda]$, with initial condition $x_0 = x_i$. Let x'_i denote the server state when it has completed service of the i th task. Then, $x'_i = 1(1 - x_i)e^{-\overline{w}S(x_i)/\tau}$. Assuming that $\overline{w}S(x_i) \leq 1/\lambda$, we get that $x_{i+1} = x'_i e^{-(1/\lambda - \overline{w}S(x_i))/\tau}$, and finally $x_{i+1} = (1 - 1(1 - x_i)e^{-\overline{w}S(x_i)/\tau})e^{(\overline{w}S(x_i) - 1/\lambda)/\tau} = (x_i - 1 + e^{\overline{w}S(x_i)/\tau}) \times e^{-(1/\lambda\tau)}$. If $\lambda, \overline{w}$ and τ are such that $x_{i+1} = x_i$, then under the trivial control policy $u(t) \equiv ON$, the server state at the beginning of all the tasks after and including the i th task will be x_i and the queue length at most 1 with the time constraint $[t_0, t_i]$. We then say that the server is at one-task equilibrium at x_i. Therefore, for a given $\lambda, \overline{w}$ and τ, the one-task equilibrium server states correspond to $x \in [0, 1]$ that satisfy $x = (x - 1 + e^{\overline{w}S(x)/\tau})e^{-(1/\lambda\tau)}$ and $S(x) \leq (\overline{w}\lambda)^{-1}$, i.e. $S(x) = (\tau/\overline{w})\log(1 - (1 - e^{1/\lambda\tau})x)$ and $S(x) \leq (\overline{w}\lambda)^{-1}$. Let us define a function $\mathcal{R}$ as

$$\mathcal{R}(x, \tau, \overline{w}, \lambda) := \frac{\tau}{\overline{w}}\log(1 - (1 - e^{1/\lambda\tau})x). \tag{4}$$

For a given $\tau > 0$ and $\lambda > 0$, define the set of one-task equilibrium server states with the time constraint $[t_0, t_i]$ as

$$x_{eq}(\tau, \overline{w}, \lambda) := \{x \in [0,1] | S(x) = \mathcal{R}(x, \tau, \overline{w}, \lambda)\}. \tag{5}$$

Note that we did not include the constraint $S(x) \leq (\bar{w}\lambda)^{-1}$ in the definition of $x_{eq}(\tau, \bar{w}, \lambda)$ in (5). This is because this constraint can be shown to be redundant as follows. Equation (4) implies that, for any $\tau > 0$ and $\lambda > 0$, $\mathcal{R}(x, \tau, \bar{w}, \lambda)$ is strictly increasing in x and hence $\mathcal{R}(x, \tau, \bar{w}, \lambda) \leq \mathcal{R}(1, \tau, \bar{w}, \lambda) = (\lambda\bar{w})^{-1}$ for all $x \in [0,1]$. Therefore, $S(x_{eq}(\tau, \bar{w}, \lambda)) = \mathcal{R}(x_{eq}(\tau, \bar{w}, \lambda), \tau, \bar{w}, \lambda) \leq (\lambda\bar{w})^{-1}$.

The strict convexity of $S(x) - \mathcal{R}(x, \tau, \bar{w}, \lambda)$ in x, which follows from the convexity assumption on $S(x)$ and the strict concavity of $\mathcal{R}$ in x from (4) within time interval $[t_1, t_i]$, implies that the cardinality of $x_{eq}(\tau, \bar{w}, \lambda)$ can take on values 0, 1, and 2. For a given $\tau > 0, \bar{w} > 0$, and $\lambda > 0$, let $x_{eq,1}(\tau, \bar{w}, \lambda)$ be the smaller element of $x_{eq}(\tau, \bar{w}, \lambda)$ if it is not empty and let $x_{eq,2}(\tau, \bar{w}, \lambda)$ be the other element if the cardinality of $x_{eq}(\tau, \bar{w}, \lambda)$ is 2. One can show that $x_{eq,1}(\tau, \bar{w})$, if it exists, is a stable equilibrium point and $x_{eq,2}(\tau, \bar{w})$, if it exists, is an unstable equilibrium point. Formally, one can show that, if $x_{eq,1}(\tau, \bar{w})$ and $x_{eq,2}(\tau, \bar{w})$ exist, then we have the following.

1) For any $\tau > 0$ and $\bar{w} > 0$, the set $(x_{eq,2}(\tau, \bar{w}),1]$ is invariant and is not in the region of attraction of $x_{eq,1}(\tau, \bar{w})$ or $x_{eq,2}(\tau, \bar{w})$.

2) There exists a $\tau^* > 0$ such that for all $\tau > \tau^*$, the set $[0, x_{eq,2}(\tau, \bar{w}))$ is invariant for all $\tau > \tau^*$. Moreover, in the limit as $\tau \to +\infty$, the set $[0, x_{eq,2}(\tau, \bar{w}))$ is the region of attraction of $x_{eq,1}(\tau, \bar{w})$.

We introduce a couple of additional definitions. For a given $\tau > 0$ and $\bar{w} > 0$, let

$$\lambda_{eq}^{max}(\tau, \bar{w})$$
$$max\{\lambda > 0 | x_{eq}(\tau, \bar{w}, \lambda) \neq \emptyset\} \tag{6}$$
$$x_{th}(\tau, \bar{w})$$
$$x_{eq,1}(\tau, \bar{w}, \lambda_{eq}^{max}(\tau, \bar{w})).$$

We now argue that the definitions in (6) with the time constraint $[t_1, t_2]$ are well posed. Consider the function $S(x) - \mathcal{R}(x, \tau, \bar{w}, \lambda)$. Since $\mathcal{R}(0, \tau, \bar{w}, \lambda) = 0$ for any $\tau > 0, \bar{w} > 0$, and $\lambda > 0$, and $S(0) > 0$, we have that $S(0) - \mathcal{R}(0, \tau, \bar{w}, \lambda) > 0$ for any $\tau > 0, \bar{w} > 0$, and $\lambda > 0$. Since $\mathcal{R}(1, \tau, \bar{w}, \lambda) = (\bar{w}\lambda)^{-1}$, $S(1) - \mathcal{R}(1, \tau, \bar{w}, \lambda) < 0$ for all $\lambda < (\bar{w}S_{max})^{-1}$. Therefore, by the continuity of $S(x) - \mathcal{R}(x, \tau, \bar{w}, \lambda)$, the set of equilibrium server states, as defined in (5), is nonempty for all $\lambda < (\bar{w}S_{max})^{-1}$. Moreover, since $\mathcal{R}(x, \tau, \bar{w}, \lambda) \leq \mathcal{R}(1, \tau, \bar{w}, \lambda) = (\bar{w}\lambda)^{-1}$ for all $x \in [0,1]$, $S(x) - \mathcal{R}(x, \tau, \bar{w}, \lambda) \geq (\bar{w}S_{min})^{-1}$ for all $x \in [0,1]$. Therefore, for all $\lambda > (\bar{w}S_{min})^{-1}$, the set of equilibrium states, as defined in (5), is empty. Hence, $\lambda_{eq}^{max}(\tau, \bar{w})$ and $x_{th}(\tau, \bar{w})$ are well defined.

In the rest of the paper, we will restrict our attention to those $\tau, \bar{w} > 0$, and $S(x)$ for which $x_{th}(\tau, \bar{w}) < 1$. Loosely speaking, this is satisfied when $S(x)$ is increasing on some interval in $[0,1]$ and the increasing part is steep enough. It is reasonable to expect this assumption to be satisfied in the context of human operators with time constraint whose performance deteriorates quickly at very high utilizations. The implications of the case when $x_{th}(\tau, \bar{w}) = 1$ are discussed briefly at appropriate places in the paper.

2)Lower Bound on the Real Time Throughput: We start by analyzing the real time throughput under a specific task release control policy. In particular, we consider the $x_{th}(\tau, \bar{w})$ threshold policy, where $x_{th}(\tau, \overline{w})$ is as defined in (6).

Theorem III.1: For any $\tau > 0$, $\overline{w} > 0$, $x_0 \in [0,1]$, $n_0 \in \mathcal{N}$, and $\lambda \le \lambda_{eq}^{max}(\tau, \overline{w})$, if $x_{th}(\tau, \overline{w}) < 1$, then we have that $\lim sup_{t \to +\infty} n_u(t, \tau, \lambda, \delta_{\overline{w}}, x_0, n_0) < +\infty$ with u being the $x_{th}(\tau, \overline{w})$ threshold policy. The proof of this result, which can be found in [20] .

3) Upper Bound on the Real Time Throughput: We now prove that the $x_{th}(\tau, \overline{w})$ threshold policy with time constraint $[t_1, t_{th}]$ is indeed maximally stabilizing by showing that no other task release control policy in $\mathcal{U}$ gives more real time throughput. Recall that a task release control policy u is maximally stabilizing within time interval $[t_0, t_\infty]$ in this setup if, for any $x_0 \in [0, 1]$, $n_0 \in \mathcal{N}$, $\lim sup_{t \to +\infty} n_u(t, \tau, \lambda, \delta_{\overline{w}}, x_0, n_0) < +\infty$ for all $sup_{u \in \mathcal{U}} \lambda_{max}(\tau, \delta_{\overline{w}}, u)$, where $\lambda_{max}(\tau, \delta_{\overline{w}}, u)$ is the throughput under policy u . We emphasize here that the allowable set of control policies $\mathcal{U}$ is pretty general and in particular, it includes, but is not limited to, threshold policies.

4. Conclusion

In this paper, we presented a real time human behavior queue framework as a formal approach to task management for human operators. Inspired by empirical laws, we considered a novel human behavior queue model for human operators, where the service times are dependent on the state of a simple underlying real time system. We studied the stability of such human behavior queues under deterministic interarrival times and real times. For homogeneous tasks, we proved that a task release control policy that releases a task to the server only when its state is below an appropriately chosen threshold value gives the maximum throughput. For heterogeneous tasks, we showed that the throughput strictly increases with the introduction of heterogeneity. The deterministic interarrival time assumption in our analysis is not binding and the results extend to the case where the interarrival times are sampled identically and independently from a common distribution having bounded variance. We also reported preliminary empirical evidence to justify the real time human behavior queue model for human operators.

We have extended the related work based on human behavior queuing system for accurate modeling of wireless IP traffic behavior through presenting a novel scheduling scheme called as Best Scheduling Algorithm (BSA). The simulation results clearly indicate that our proposed scheduling algorithm outperforms the traditional scheduling schemes such as priority and round-robin. The BSA provides a preferential treatment to real time applications by offering a very low delay but at the same time, this preference is not up to that extent that generic data applications are starving for bandwidth. In our future work, we are intending to explore the possibility of practical implementation of proposed BSA in different 4G wireless networks.

References

[1] G. Armitage, "Quality of Service in IP Networks", MTP, pps 105-138, 2004.

[2] K. Savla, T. Temple, and E. Frazzoli, Human-in-the-loop vehicle routing policies for dynamic environments, Proceeding. IEEE Conf. Decision Control, pp. 1145¡V1150, 2008.

[3] J. H. Dshalalow, Ed., BQueueing systems with state dependent parameters,[in Frontiers in Queuing Models and Applications in Science and Engineering. Boca Raton, FL: CRC Press, 1997.

[4] R. Bekker and S. C. Borst, BOptimal admission control in queues with workload-dependent service rates,[Probab. Eng. Inf. Sci., vol. 20, pp. 543¡V570, 2006.

[5] B. Tsybakov and N. D. Georganas, "Self-Similar traffic and upper bounds to buffer overflow in ATM queue", Performance Evaluation, 36, pps. 57-80, 1998.

[6] R. Addie, M. Zukerman and T. Naeme, "Fractal traffic: measurements, modeling and performance evaluation", Proceeding IEEE INFOCOMM 95, pp. 977-984, 1995.

[7] M. Zukerman et al, "Analytical Performance Evaluation of a Two Class DiftServ Link", IEEE ICS, vol. 1, pp. 373-377, 25-28 Nov. 2002.

[8] J. Zhang, "Performance study of Markov modulated fluid flow models with priority traffic", in Proc. IEEE INFOCOM 93, San Francisco, CA, pps 10-17, Mar. 30-Apr. 1, 1993.

[9] Do. Young. Eun and Ness. B. Shroff, "A measurementanalytical approach for QoS estimation in a network based on dominant time scale" in IEEE/ACM Trans. on Networking, vol. II, No. 2, pps. 222-235, April 2003.

[10] A. I. Elwalid and D. Mitra, "Fluid Models for analysis and design of statistical multiplexing with loss priorities on multiple classes of bursty traffic" in Proc. IEEE INFOCOM 92, Florence, Italy, pps 415-425, 1992.

[11] G. L. Choudhury, K. K. Leung and W. Whitt, "An inversion algorithm to compute blocking probabilities in loss networks with state-dependant rates", IEEE/ACM Transactions on Networking, vol. 3. pp. 585-601, 1995.

[12] C. F. Chou et ai, "Low Latency and efficient packet scheduling for streaming applications" IEEE International Conforence on Communications, Vol. 4, pp. 1963-1967, 20-24 June, 2004.

[13] J. M. Chung, H. M. Soo, "Analysis of Non Preemptive Priority Queueing of MPLS networks with Bulk Arrivals", IEEE MWSCAS, vol. 3. pps 81-84, 4-7 Aug. 2002.

[14] M. Iftikhar et aI, "SLAs parameter negotiation between heterogeneous 4G wireless network operators", Elseveir Journal of Pervasive and Mobile Computing, vol. 7, issue 5, pps. 525-544, October 2011.

[15] M. Iftikhar et aI, "Towards the formation of comprehensive SLAs between heterogeneous wireless DiffServ domains", Springer Journal of Telecommunication Systems, 42: 179-199, 2009.

[16] M. L. Cummings and C. E. Nehme, BModeling the impact of workload in network centric supervisory control settings, Proceeding. 2nd Annu. Sustaining Performance Under Stress Symp., College Park, MD, Feb. 2009.

[17] P. A. Hancock and N. Meshkati, Eds., Human Mental Workload, vol. 52, Advances in Psychology. Amsterdam, The Netherlands: Elsevier Science, 1988.

[18] R. M. Yerkes and J. D. Dodson, BThe relation of strength of stimulus to rapidity of habit-formation, J. Comparative Neurol. Psychol., vol. 18, pp. 459¡V482, 1908.

[19] K. Savla, C. Nehme, T. Temple, and E. Frazzoli, BEfficient routing of multiple vehicles for human-supervised services in a dynamic environment, Proceeding AIAA Conf. Guid. Navig. Control, Honolulu, HI, 2008, Paper AIAA 2008-6841.

[20] K. Savla and E. Frazzoli, Maximally stabilizing admission control policy for a dynamical queue, IEEE Trans. Autom. Control, vol. 55, no. 11, pp. 2655¡V2660, Nov. 2010.

9

SIMULATION STUDIES ON AN ENERGY EFFICIENT MULTIPATH ROUTING PROTOCOL USING DIRECTIONAL ANTENNAS FOR MANETs

Sandhya Chilukuri[1], Rinki Sharma[2], Deepali. R. Borade[3] and Govind R. Kadambi[4]

Department of Computer Engineering, M. S. Ramaiah School of Advanced Studies, Bengaluru, India
reachsandhyach@gmail.com;govind@msrsas.org;rinki@msrsas.org;
borade.deepali21@gmail.com

Abstract

The paper proposes the development of an energy efficient multipath routing protocol with directional antenna for MANET as an optimization task as well as a multidisciplinary entity. A comprehensive analysis to link all the multi-disciplinary viewpoints involved in the development of desired multipath routing protocol with requisite technical details is presented in this paper. A simple and elegant mathematical formulation for the analysis of relative improvement of the performance metrics of ad-hoc networks with omnidirectional and directional antenna is presented. Through extensive numerical simulations, the multi-dimensional desirable performance attributes of wireless link such as improved range, improved RSS, reduced RF transmit power and consequent reduced consumption of battery power have been analyzed keeping the directional gain of the antenna as a variable parameter. Development or modification of a protocol with concurrent focus on multipath routing with optimization of the battery energy is a significant step to increase the life time of MANET without recharging. Selection of the energy efficient path amongst several alternative ones is of paramount significance in the evaluation of overall performance of MANET system. A formulation to compute the required Battery Energy taking into account the data pertaining to the power efficiency of the associated transceiver design as well as the specified link performance parameters is also discussed. Through a case study involving the specifications of a typical transceiver operating in the 2.4 GHz band, the desirable impact of higher gain of a directional antenna in the reduction of RF transmitter power is illustrated. The consequential reduced battery power consumption while still retaining the specified performance parameters of the ad-hoc network like range and Received Signal Strength (RSS) is also demonstrated. This paper also addresses the importance of alignment of beam peaks of directional antennas of a link and the quantification of additional RF power in lieu of Beam Pointing Angle (BPA) error in ad-hoc network. The profile of improved range with directional gain as an independent variable exhibits much sharper feature than an exponential function. The relationship between improvement in the RSS and higher directional gain bears linear characteristics and typical results reveal that for a dB increase in gain ratio, the corresponding improvement in RSS is 2 dBm. Results of Proposed Routing protocol simulations with built in hop count reduction feature reveal energy saving of 62.11% for a typical MANET scenario of 25 nodes, 5 MB data and data rate of 40 Kbps.

Keywords

Ad-Hoc Network, Mobile Ad-Hoc Network, Energy Efficient, Directional Antenna, Routing Protocol,

1. INTRODUCTION

Multi Hop Wireless Network has already emerged and continues to be a topic of hot research pursuit. An ad-hoc network is established with multiple mobile nodes, coming in range of each other and exchanging data without an access point. The ability of participating nodes to move around gives rise to Mobile Ad hoc Networks (MANETs). Such networks are very useful for military applications, emergency and rescue operations, health care, home networking and other commercial and educational applications. In a MANET, nodes are highly interdependent for exchange of information with each other. The nodes must cooperate to routing and other services, making it crucial to maximize the network lifetime. The participating nodes in a MANET (laptops, PDA's and sensors) have limited battery power; therefore energy efficient communication approach is critical for the longevity and efficiency of the network. To support energy-efficient communication in an Ad-hoc network, it is important to apply energy-efficient design at multiple layers of the network protocol stack. To realize an energy efficient ad-hoc network is to minimize the RF transmit power for conservation of battery power. In view of this, this paper emphasizes the application of directional antenna to realize the reduction in transmitter power and yet maintaining the desired link performance or network performance.

In this paper, we concentrate on the physical layer techniques used for energy-efficient communication in an attempt to illustrate the enhanced performance of Ad-hoc networks with directional antenna. Analysis to link the significant design parameters of directional antenna to wireless link performance is of practical significance to appreciate the system design considerations of ad-hoc networks. Through simple and elegant mathematical formulation, relative improvement of the performance metrics of ad-hoc network is analyzed when high gain directional antenna is replaced with conventional low gain omni directional antenna. Resource utilization particularly that of electric power of battery and the associated energy is of paramount importance since it has significant say in the outage condition of a wireless link or network. Through a case study involving the specifications of a typical transceiver operating in the 2.4GHz band, the desirable impact of higher gain of a directional antenna in the reduction of RF transmitter power is illustrated. This in turn results in the reduced consumption of battery power, while still retaining the specified performance parameters of the ad-hoc network such as range and Received Signal Strength (RSS). Although, directional antenna offers potentially many desirable performance improvement attributes, it is also associated with additional constraint or requirement of precise or accurate alignment of the directive or main beam of the Transmitter and Receiver. The mismatch of angular alignment of the Transmitter and Receiver results in Beam Pointing Angle (BPA) error, which leads to the degradation in the range. If one desires to regain the desirable ideal network performance despite the presence of BPA error, it would call for additional RF Transmitter power ultimately culminating in the consumption of extra power of the battery that support the RF operation. This paper also facilitates the quantification of additional RF power in lieu of BPA error in ad-hoc network.

Development or Modification of a protocol with concurrent focus on multipath routing with optimization of the battery energy is a significant step to increase the life time of MANETs without recharging. Selection of the energy efficient path amongst several alternative ones is of paramount significance in the evaluation of overall performance of MANET system. The desired protocol development is an optimization exercise involving multidisciplinary aspects of Antenna Engineering, Microwave Link Budget to treat intermediate wireless links, Algorithms to determine the most efficient path for reliable connectivity between the source and destination nodes, comprehensive relation between the RF power and the pertinent Battery Power of the transmit nodes involving conventional electrical power engineering basics as well as the choice and optimization of route metrics to arrive at decision logic for the path selection.

2. COMPREHENSIVE SYSTEM VIEW OF MANET

Figure 1 depicts a comprehensive System View of energy efficient multipath MANET routing protocol. Antenna or microwave link budget includes the concept of gain, link data, battery energy, range, RF power, RSS which are interrelated with each other. A perfect combination of them will help in conserve energy while routing.

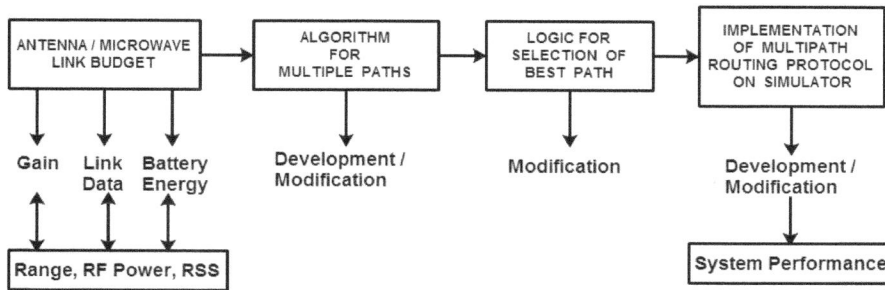

Figure 1: Comprehensive System View of Energy Efficient Multipath MANET Routing Protocol

From the system design perspective, this paper addresses basically two broad system entities namely antenna and its relevance to energy efficiency considerations in MANET.

3. DIRECTIONAL ANTENNA IN MOBILE AD HOC NETWORKS (MANETs)

A representative generic MANET with three mobile nodes is shown in Figure 2 with the dotted circles, implying the communication range of each node. Each of the nodes is assumed to be design configured with a RF transceiver as their network interface. If Node A wishes to communicate with Node C and if Node C is not in communication range of Node A, Node B will serve as an intermediate node resulting in 2 hop communication.

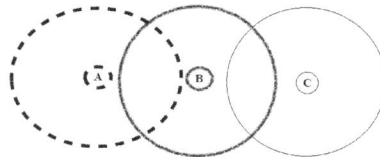

Figure 2. Multi Hop Communication in MANET

A transceiver of a MANET is configured with an antenna which will serve as its RF front end to establish a wireless communication link with its intermediate (participating) nodes. Basically, the antenna can be of two types namely omnidirectional and directional. Figure 3 shows the difference between radiation patterns of directional and omnidirectional antenna. It can be seen that the radiation pattern of omnidirectional antenna exhibits uniform angular distribution while the radiation pattern of directional antenna is concentrated in the preferred angle (direction) of communication. In Figure 3, S and D denote the source and destination nodes respectively. As illustrated in Figure 3, the communication range exhibited by directional antenna will be higher than an omnidirectional antenna. In view of this, it is implicit that the transceivers of nodes in Figure 2 are associated with omnidirectional antennas. In the past, omnidirectional antennas did find extensive applicability in MANET system design. With the rapid advancement of antenna technology conjunctured with miniaturization in size and its proven system utility in subscriber end of cellular communication, increasingly one finds rapid progression of directional antenna in ad-hoc networks/MANETs. The characterization of directional antenna to realize performance improvements in ad-hoc networks/MANETs wherein mobility is a key factor causing potentially dynamic and vast variations in communication range, constitutes the core theme of this paper.

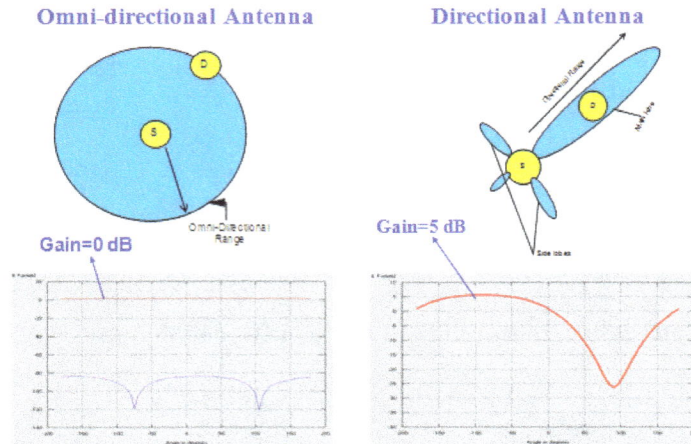

Figure 3. Radiation Patterns of Omnidirectional and Directional Antennas

4. Energy Saving Techniques at Routing Layer in MANETs

Energy efficient routing protocols can be mainly categorised into two types. One category selects the path that has maximum total energy, while the other category of protocols selects the paths with nodes having residual energy greater than energy required to transfer data.

The protocols that select the path with maximum energy ensure that when a packet is transmitted from source to destination, there is enough energy available for the exchange of data. While the other category of protocols which selects the path with nodes having residual energy greater than energy required to transfer data and ensures that each node has enough energy on the selected path. However, if the path selected does not minimize the energy needed to transmit a packet from its source to its destination, the network lifetime may not be maximized. Therefore, the usage of directional antenna helps in reducing the number of hops to conserve the battery energy and in turn helps in maximizing the network lifetime.

5. RELATED WORK ON DIRECTIONAL ANTENNA AND MULTIPATH ROUTING IN MANETs

Most of the papers available in literature propose MAC layer or network layer protocols for ad-hoc networks, using directional antenna. Su Yi et.al [1] provide a good insight on the improvement in capacity of wireless network with the usage of directional antenna. This paper also provides results for throughput improvement with combination of omnidirectional and directional antennas in Transmit (Source) and Receive (Destination) nodes. In [2] Ram Ramnathan provides a good insight on beamforming antennas, and their use in ad-hoc networks. This paper also provides a rough comparison of relative lesser interference potential with: only beamforming, only power control and when both beamforming and power control are used together. The research in [3,4,5] suggests the use of directional antennas for increasing the throughput and enhancing the performance of wireless networks.

However, the above cited research papers do not address a detailed formulation to link the relative improvement in Received Signal Strength (RSS), range and reduction in transmit power when high gain or moderate gain directional antenna replaces the conventional low gain omnidirectional antenna in a MANET system. This paper illustrates the potential of directional antenna to ensure the performance improvement of the communication link relative to omnidirectional antenna. Further, with the directional antennas establishing the link, it is all the more necessary to ensure that the direction of beam (peak gain) of the antenna of the transmit node is aligned with the direction of peak gain of the antenna of the receive node. In case of mismatch in antenna beam pointing angle between the two communicating nodes, there shall be a consequent degradation in the performance of communication link. This paper provides a quantitative analysis of the degradation in performance of the communication link in the presence of misalignment (angular error) between the beam pointing angles of transmit and receive nodes. An estimate of the required increase in the transmit power of the link to avoid the performance degradation and to maintain link quality as with ideal zero beam pointing error case is also provided.

The initial design and evaluation of two techniques for routing improvement using directional antennas in mobile ad hoc networks are discussed in [6]. In the first step directional antenna is used to transmit selected packets over a longer distance. Relatively more number of packets is transmitted over a shorter distance. The adaptive combination of two is called as bridging permanent network partitions. In a subsequent step, network without permanent partitions, directional antenna is used to repair routes in use, by using the capability of directional antenna to transmit packets over longer distance. In [7] the authors have proposed an energy-efficient, distributed, scalable and localized multipath search algorithm to discover multiple node-disjoint paths between the sink and source nodes for wireless sensor networks. A reactive source routing protocol referred to as Multipath Directional Antenna ad-hoc Routing (MDAR) is proposed in [8]. A distinctive feature of MDAR is that the routing table records multiple choices of routes to each destination, so that when one route encounters busy channel, an alternative route can be selected immediately.

6. ANALYSIS AND RELATIVE COMPARISON OF LINK PERFORMANCE WITH DIRECTIONAL AND OMNIDIRECTIONAL ANTENNAS

From the joint perspectives of antenna engineering and its relevance to system performance of the links established through participating nodes, the communication range, RSS, Transmitter power to maintain the link performance and resource utilisation particularly with respect to Battery power emerge as more critical parameters. This section provides extensive analysis of the enhancement of the parameters listed above with the use of directional antenna when compared to that of omnidirectional antenna.

6.1. Improvement in Range due to Gain of Directional Antenna

Friis transmission formula is used in wireless communication to calculate the power received by the antenna at the receiver section under idealized conditions. Friis transmission formula is defined through

$$P_r = P_t \times G_t \times G_r \times \left(\frac{\lambda}{4\pi R}\right)^2 \qquad \text{--------------------(1)}$$

Where, P_t, P_r, G_t, G_r, R and λ represent Transmit Power, Receiver Power, Transmit antenna gain, Receive antenna gain, Range and Wavelength respectively.

Let, R_{omni}, RSS_{omni}, P_{Tomni}, λ, G_{Tomni} and G_{Romni} represent Range of Omnidirectional antenna, Transmit power with omnidirectional antenna, Wavelength in meters, Gain of Transmit omnidirectional antenna and Gain of receive omnidirectional antenna respectively.

Then we have,

$$RSS_{omni} - P_{T\,omni} \times G_{T\,omni} \times G_{R\,omni} \times \frac{\lambda^2}{4^2 \pi^2 R_{omni}^2}$$

$$R_{omni}^2 = P_{T\,omni} \times G_{T\,omni} \times G_{R\,omni} \times \frac{\lambda^2}{4^2 \pi^2 RSS_{omni}}$$

$$R_{omni} = \frac{\lambda}{4\pi} \sqrt{\frac{P_{T\,omni}\,G_{T\,omni}\,G_{R\,omni}}{RSS_{omni}}} \qquad \text{-------------------(2)}$$

The increase in the communication range ΔR due to directional antenna can be related to the term directional gain ratio g_r, where $g_r = G_{DA}/G_{OMNI}$

$$\Delta R = R_{omni}\,(g_r - 1) \qquad \text{--------------------(3a)}$$

Effective range $R_{effective} = R_{omni} + \Delta R$ $\qquad$ --------------------(3b)

In Equation (3), g_r involves G_{DA} and G_{omni} where,

G_{DA} = Gain of Directional Antenna ($G_{DA} = G_{TDA} = G_{RDA}$)

G_{omni} = Gain of Omnidirectional Antenna ($G_{omni} = G_{T\,omni} = G_{R\,omni}$)

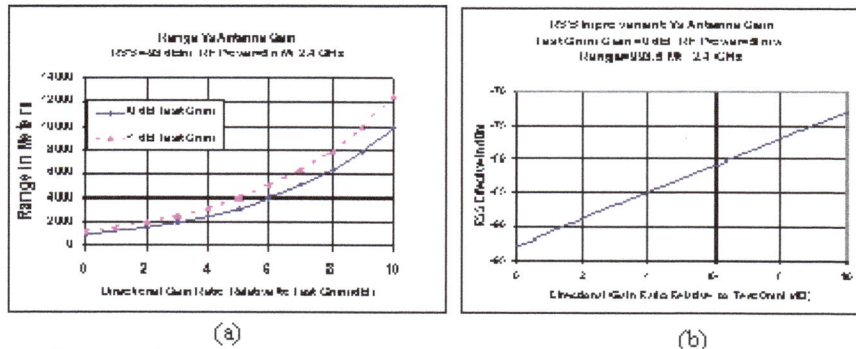

Figure 4. Influence of Directional Antenna Gain on Communication Range and Improvement of RSS effective with Directional Gain

Figure 4(a) illustrates the improvement in the communication range solely due to the gain of the directional antenna relative to omnidirectional antenna keeping the transmit power P_T =5mW and RSS=-93dBm to be the same in both cases of directional as well as omnidirectional antennas with the link operating at 2.4 GHz. The profile of increased gain variation exhibits much sharper feature than an exponential function.

6.2. Improvement in Received Signal Strength (RSS) due to Directional Antenna

If the parameter P_r in Equation 1 is considered as RSS_{omni}, with G_{Tomni} and G_{Romni} as omnidirectional antenna gain, then we have,

$$RSS_{omni} = P_{T\,omni} \times G_{T\,omni} \times G_{R\,omni} \times \left(\frac{\lambda}{4\pi R_{omni}}\right)^2 \quad \text{------------(4)}$$

The improvement in the RSS with the use of directional antenna can be written as,

$$\Delta_{RSS} = RSS_{omni}\left[(g_r)^2 - 1\right] \quad \text{------------(5)}$$

The effective RSS at the receiver of the link with the directional antennas replacing the omnidirectional antennas is

$$RSS_{effective} = RSS_{omni} + \Delta_{RSS} \quad \text{------------(6)}$$

Figure 4(b) depicts the variation of realizable improved $RSS_{effective}$ as a function of the directional gain ratio. Apart from linear nature of variation, the result depicted in Figure 4(b) also reveals that for a dB increase in gain ratio, the corresponding improvement in $RSS_{effective}$ is 2dBm.

In this section, we show the reduction in required transmission power with the use of directional antenna, when compared to that required for omnidirectional antenna. Based on the Friis Transmission formula given by Equation 1, we have,

$$P_{T\,omni} = \frac{RSS_{omni}\,4^2\pi^2 R_{omni}^2}{G_{T\,omni}\,G_{R\,omni}\,\lambda^2} \quad \text{---------------(7a)}$$

The reduction in the Transmitter power ΔP_T because of higher gain of directional antenna can be expressed as

$$\Delta P_T = P_{T\,omni}\left[1 - \left(1/g_r\right)^2\right] \quad \text{----------------(7b)}$$

The definition of Gain Ratio remains the same as in earlier sections. The actual transmit power required by the directional antenna to retain the range and RSS_{omni} of the omnidirectional antenna is given by

$$P_{T\,effective} = P_{T\,omni} - \Delta P_T \quad \text{------------------(8)}$$

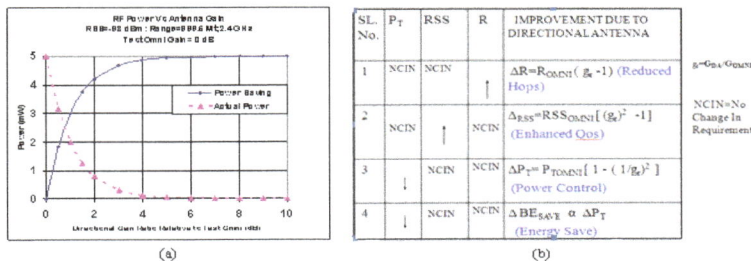

(a)

SL. No.	P_T	RSS	R	IMPROVEMENT DUE TO DIRECTIONAL ANTENNA	
1	NCIN	NCIN	↑	$\Delta R = R_{OMNI}(g_r - 1)$ (Reduced Hops)	$g_r = G_{D4}/G_{OMNI}$
2	NCIN	↑	NCIN	$\Delta_{RSS} = RSS_{OMNI}[(g_r)^2 - 1]$ (Enhanced Qos)	NCIN=No Change In Requirement
3	↓	NCIN	NCIN	$\Delta P_T = P_{TOMNI}[1 - (1/g_r)^2]$ (Power Control)	
4	↓	NCIN	NCIN	$\Delta BE_{SAVE} \propto \Delta P_T$ (Energy Save)	

(b)

Figure 5. Reduction in RF Transmitter Power with Directional Gain and Multifaceted Advantages of Directional Antenna

Figure 5(a) shows the reduction in the RF Transmitter power with the use of directional antenna keeping the RSS as well as range parameters the same in both the directional and omnidirectional cases. Drastic reduction in transmitter power is evident with smaller initial increase in directional gain and then exhibiting asymptotic nature with higher directional gain.

6.3. Multidimensional Versatility of Directional Antenna

Figure. 5(b) summarizes the highlights the advantages of using directional antenna. In Figure. 5(b), ΔBE_{Save} denotes the reduction in the energy of battery with the directional antenna to retain or maintain the link performance as in the case of omnidirectional antenna.

With increased communication range (R), the number of hops required for data exchange in an ad-hoc network can be reduced thus reducing the end-to-end latency. Higher RSS increases the signal quality at the receiver, thus providing enhancement in Quality of Service (QoS). With the reduction in transmitter power, the battery energy consumed at each node decreases, thus providing longer battery life, which further increases the network lifetime. Power control also reduces interference among neighboring nodes thus increasing the efficiency of the network.

6.4 Estimation of Battery Energy for Given Data Transmission

Research publications [9,10,11,12] address the issue of need and challenges in the realization of energy efficient ad-hoc/MANET. However, a formulation or a detailed formulation with clearly illustrated steps to compute the battery energy required for a specified data transmission seems to have not been addressed in the literature even though this is very significant from system design perspective. This section summarizes the detailed steps through which one could compute the required battery energy and the formulation shall take into account the data pertaining to the power efficiency of the associated transceiver design as well as the specified link performance parameters.

From the classical relationship between Power, Energy and Time, one can draw analogy to relate the requirement of Battery Power (P_{Bat}) and Battery Energy (BE_{Data}) for transmission of given data.

$$BE_{Data} = P_{Bat} x T_{ON} \qquad\qquad \text{------------------(9)}$$

Where T_{ON} = Transmitter ON Time.

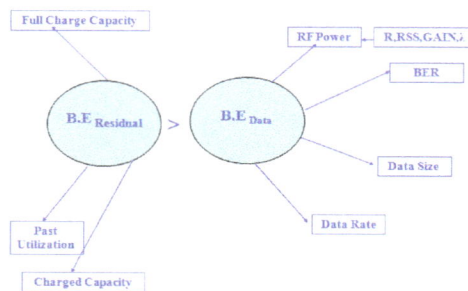

Figure 6. Requirement of Battery Energy for Data Transmission

T_{ON} is dependent on the size of data, data rate as well as channel condition between the Transmit and Receive ends of the link. At an instant of time for data transmission, the following inequality should be satisfied.

$$BE_{Residual} > BE_{Data} \qquad\qquad \text{---------------(10)}$$

Where, $BE_{Residual}$ denotes the available Battery Energy at that instant for operation. Many research publications dwell with energy efficient routing protocols in ad-hoc/MANET. However, it is difficult to find a formulation to estimate or compute BE_{Data} that has a generic or empirical appeal and which facilitates a better appreciation from system design considerations. In this section, a generic procedure is outlined to systematically compute the BE_{Data}. As illustrated in Figure 6, the Battery Energy is dependent on communication range (R), RF power of transmitter, gain of the transmitting as well as receiving antennas, the expected or threshold RSS at the receive node and wavelength.

For a specified range, RSS, Gain of Transmit and Receive antennas and wavelength, the RF Transmit Power can be computed using equation (11):

$$P_T = \frac{RSS\ 4^2\ \pi^2 R^2}{G_T\ G_n \lambda^2} \qquad\qquad \text{---------------(11)}$$

The relationship between P_T and Battery Power P_{Bat} is very specific to the RF design of the Transceiver module of the ad-hoc/MANETs as well as the type of battery used in it. However, the data sheet of RF transceiver usually provides the relationship between P_T and P_{Bat} in tabular form. Either a look up table or an algebraic expression deduced from the tabulated data is required to estimate the P_{Bat} for a specific PT determined through Equation (11). Such recourse of computing the Battery Power P_{Bat} shall further facilitate built in provision for computation of Battery Energy BE_{Data} through Equation (9). Figure 7 illustrates the least square curve fitting equation obtained to relate the RF Power P_T and Battery Power P_{Bat} for a typical RF transceiver widely used in 2.4 GHz band. Satisfactory correlation between the results of derived functional relation and the reference data is noticed in Figure 7.

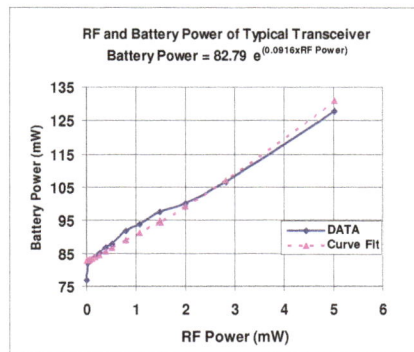

Figure 7. Derivation of Functional Relation between RF and Battery Power

6.5 Effect of Beam Pointing Angle Error of Directional Antenna on Network Performance

In order to retain all the network performance enhancements highlighted thus for, one has to also address the importance of alignment of direction of peak gain of the transmitting beam of the source node with the corresponding receiving beam of the destination node. Misalignment of the beam peaks of the source and destination nodes result in a parameter defined as a Beam Pointing Angle (BPA) Error or Beam Pointing Error (BPE) as shown in Figure. 8. In many of the aperture antennas with moderate gain, the shape of the radiation pattern up to -10 dB points is assumed to follow functional distribution $e^{-p\theta^2}$.

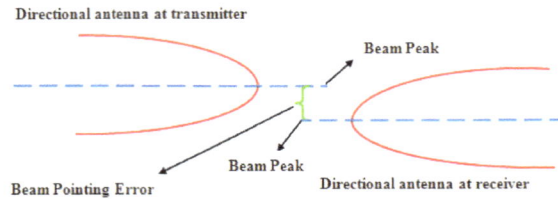

Figure 8. Illustration of Beam Pointing Angle Error of Directional Antenna

The value of the constant 'p' can be derived with the plot of the radiation pattern at say -3 dB or -10 dB points relative to the beam maximum angle. This procedure has been adopted to analyse the effect of BPA or BPE on the link performance metrics such as range and the RF power requirement as well as Battery Power. The degradation in the range performance as a function of BPA is shown in Figure 9(a). As expected, the degradation becomes more pronounced with increase in BPA.

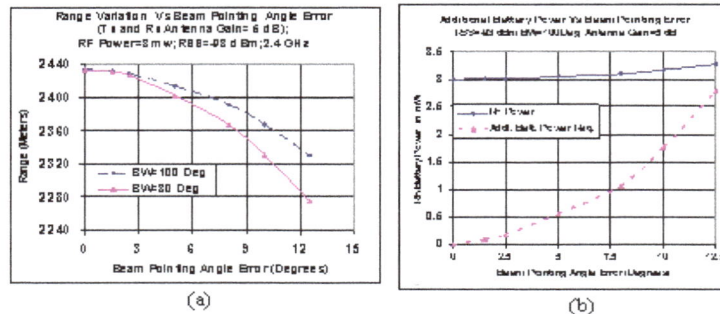

Figure 9. Influence of Beam Pointing Angle Error on Network Performance and Effect of Beam Pointing Angle Error on RF and Battery Power

If one desires to regain the ideal link performance despite the presence of BPA or BPE, it can be realized only at the expense of increased RF transmit power. This in turn would result in additional consumption of battery power resulting in reduced operational time of the link. Figure 9(b) explains the undesirable influence of BPA on the link transmitter power and which in turn has a negative effect on the required battery power to sustain the link performance.

7. Energy Efficient Multi Path Routing Protocol

The Energy Efficient Protocol proposed in this paper lists all possible paths from source to the destination with no repetition and avoiding loop formation in path computation. Once the multiple paths are computed, next step is to decide which among them is the best from energy efficiency consideration. To arrive at that decision, the protocol ranks each path based on Hop count, battery energy required to send the data and residue of residual energy of each path. From among all the multiple paths, the path with highest weightage which corresponds to the most energy efficient is selected.

7.1 Hop Count Minimization

Once the most energy efficient path is chosen based on Decision Weights, further optimization of the number of hops is the next step. Ideal requirement is that the data transfer be accomplished with a single hop irrespective of the distance between the source and the destination nodes. However, the power handling capacity of node, gain of the antenna associated , desired RSS at the Receiver and Wavelength determine the maximum communication range that is realizable and is calculated using the formula

$$\max\ Dir = \left[\frac{\lambda}{4\pi}\right]\sqrt{\frac{P_{TDir} \times G_{TDir} \times G_{RDir}}{RSS_{Dir}}} \qquad \ldots\ldots\ldots\ldots(12)$$

For $G_{TDir} = G_{RDir}$ = 5 dB, P_{Tdir} = 5 mW, RSS_{Dir} = -93 dBm and λ = 0.125 m, the maximum range is max Dir=3141.87 m. The disadvantage of selected optimized path when the size of the network is large is that the distance between nodes is also very large. It is also associated with a relatively higher number of hop counts which in turn may not ensure error free communication. To overcome this, Proposed Protocol has a feature of hop count minimization using maximum range of directional antenna. Using directional max distance as a parameter, the protocol searches the possible minimum number of hops required to cover the already chosen energy efficient multipath.

From Figure 10, it can be seen that initial path selected by the protocol has 9 hops between yjr source node 11 and destination node 24. Using directional max range that is 3141.87m, the distance between source and the destination nodes is considered to check whether it is less than directional 'max Dir', so that possibility of one hop communication can be ascertained. But in the scenario under consideration, the distance between the source-destination nodes is 3824.3 m which is greater than directional max range implying the non-feasibility of one hop communication (Figure 10).

```
Path Selected   11      13      14      17      18      19      20      21      22      24
*******************HOP_ARRAY***************
        11      21      22      24
*******************HOP ARRAY***************
```

Figure 10. Multi Hop Communication between Source Node 11 and Destination Node 24

If one hop communication between source and destination nodes is not possible, then .the distance between source node 11 and the node 22 which is next closest node to the destination node and also furthest from the source node next only to the destination node is checked. The range between the source node 11 and the node 22 is 3259.6 m which is still higher than the directional max range. Recursively, the next node from the destination end that is farthest to the source node is chosen and the range between these is checked. As shown in Figure 11, distance between source 11 and node 21 is verified and its value of 3000 m is less than directional max range. This means that node 21 is within one hop communication range of directional antenna of source 11. So all the nodes in the chosen energy efficient multipath between source node 11 and node 21 are skipped and the final hop count optimized path is of 3 hops and this is elucidated from Figure 11. In the chosen scenario, the number of hops of the selected path without refined hop count optimization is 9 which has been reduced to 3 hops which ensures error free and energy efficient communication when compared to communication through selected path.

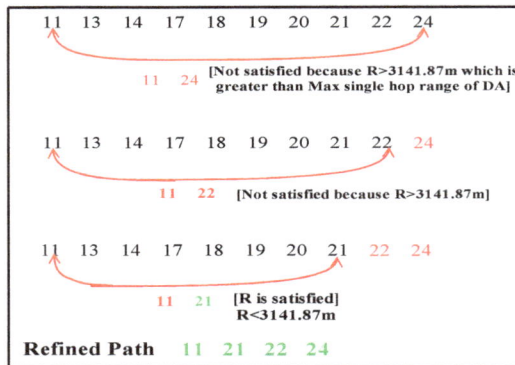

Figure 11. Pictorial Representation of Hop Count Minimization

7.2 Energy Conserved Using Directional Antenna Hop Count Minimization

Hop count minimization feature of the proposed protocol is very useful when the network size is large. From Table 1, it is seen that energy utilized by selected multiple path with 9 hops is 94.83 Joules where as the energy used by hop count optimized path with 3 hops is 36.72 Joules. In these computations, the data size of 5 MB and data rate of 40 kbps have bee assumed. Hence the energy save is 58.11 Joules which amounts to 62.11%. Thus the stated principal theme of development and demonstration of energy efficient multi path routing protocol using directional antenna for MANET featured with reduction of number of hop counts potentially assures not only energy saving but also an error free communication.

Table 1. Energy Utilization and Conservation between Two Paths

Paths	BE Used to Send Data(J)	Hop Count
11 13 14 17 18 19 20 21 22 24	94.83	9
11 21 22 24	36.72	3
Energy Saved	58.11	6

62.11 % Energy Saved

8. Results

Figure 12 displays the QualNet simulation showing the placement of 25 nodes in the network(Give some additional details).

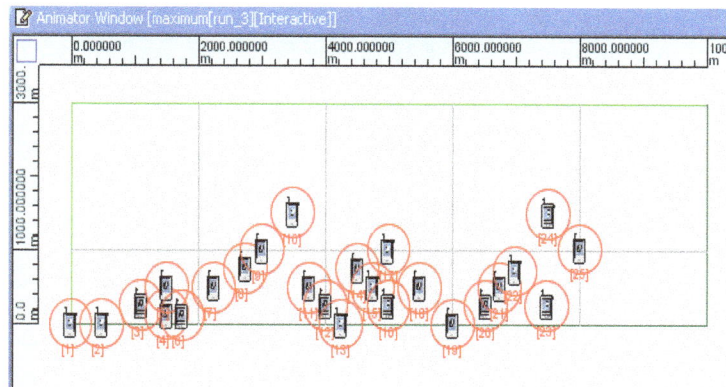

Figure 12. QualNet Animator (B)

8.1 Performance Analysis

A. Results of Packet Delivery Ratio Results of Scenario with 25 nodes

Figure 13 shows the error free data transmission from the source node 11 to destination node 24 using the refined final path following the hop count minimization feature of the proposed protocol.

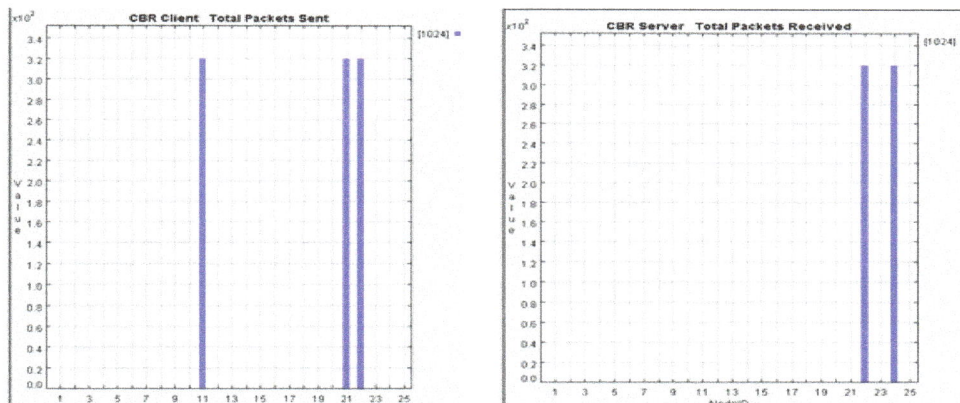

Figure 13. Packet Sent and Received

When the path for data communication contains more number of hops, usually error free communication is not expected/ensured. The combination of directional antenna, Energy Efficient multipath selection and further refinement of the selected path with optimization of hop counts has resulted not only an enormous saving of energy as shown Table 1 but also an error free transmission,

8. 2 Analysis of Throughput

Figure 12 is a typical depiction of relatively larger size of network of 25 nodes. Throughput is an important performance metric of any wireless network. As can be seen from the results of Figure 14, the total number of 79 Kilo Bytes transmitted by the sender node 11 has been received by the first intermediary nodes 21.

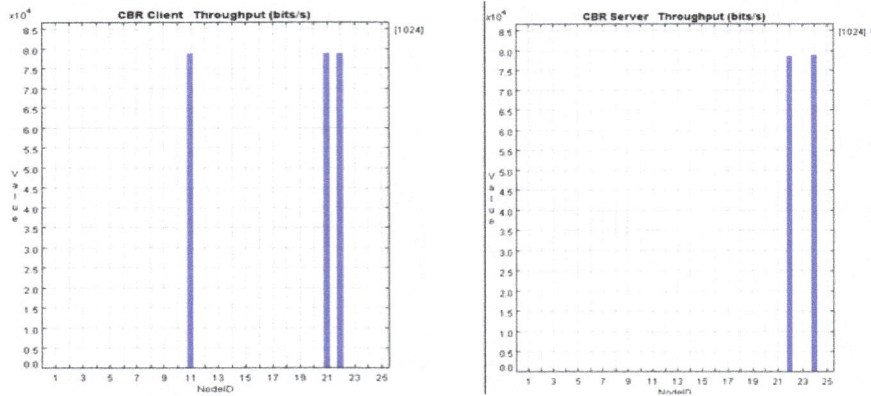

Figure 14. Throughput

Subsequently the transmission of it by node 21 has been received by the second intermediary node 22 also and ultimately culminating in the reception of all the 79 Kilobytes by the destination node 24. Figure 14 amply demonstrates the effectiveness of the energy efficient multipath routing protocol using directional antenna proposed in this paper.

8.3 Performance Analysis of End to End Delay

A. Scenario with 25 Nodes (9 Hops)

The delay (elapsed time between the transmission and reception at the successive nodes) strongly depends on the distance between the transmit and receive nodes. The time elapsed between the communication starting with the transmission at the source node and the reception at the destination node is the algebraic summation of the delays caused at each hop along the selected multi path. The Figure 15 displays the amount of time elapsed between the transmit and reception processes at the successive nodes along the selected path comprising 9 hops. The end to end delay encountered by the individual hops varies from a minimum of 5 ms to a maximum of 8.6 ms.

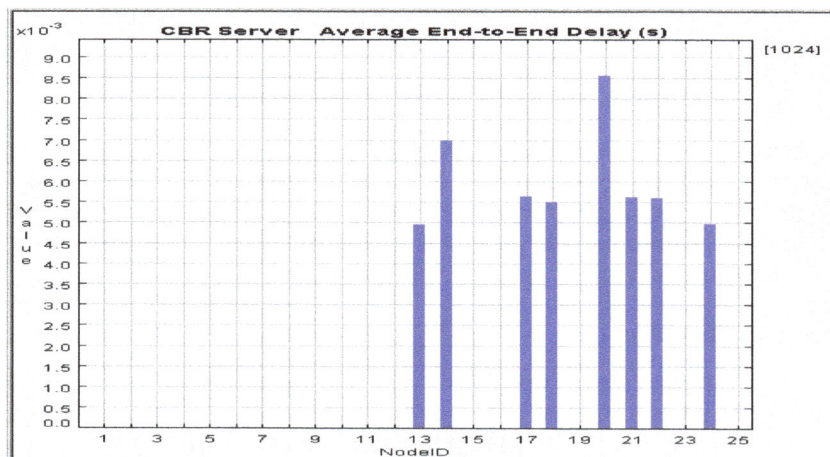

Figure 15. End To End Delay of Selected Path

B. Scenario with 25 Nodes (3 Hops)

The Figure 16 intends to illustrate the advantage of the hop count optimization feature of the proposed protocol. While the results of Figure 15 pertain to the energy efficient multipath chosen by the decision weight of the analysis for 25 node scenario with 9 hops, the Figure 16 refers to the same scenario but with the number hops being reduced to 3. The results of Figure 16 spring no surprises since it is on expected lines in the sense that smaller the hop counts, lesser the spread in the variation of end to end delay. A nearly uniform and a relatively lesser end to end delay of about 3.4s is observed along the refined energy efficient multipath.

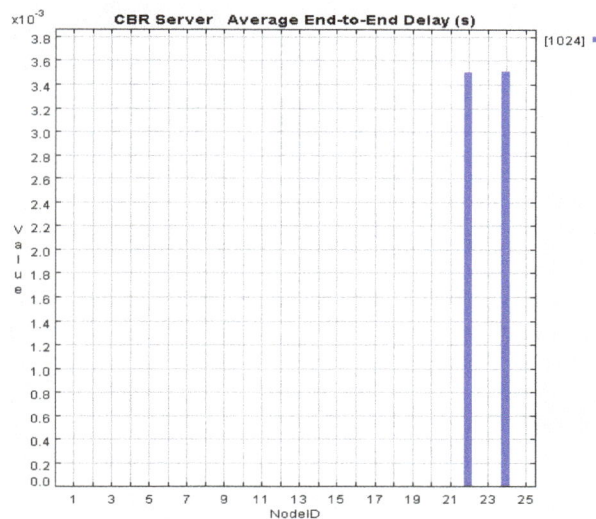

Figure 16. End To End Delay of Hop Count Optimized Path

8.4   Performance Analysis of Jitter

A. Scenario with 25 Nodes (9 Hops)

Jitter is also one of the parameters to be considered in the performance evaluation of a Wireless Communication system such as MANET. It is defined as variation in time between the packets arriving at the destination. The variation can be attributed to one or more of the followings; Network Congestion, end to end to delay or changes in the route.

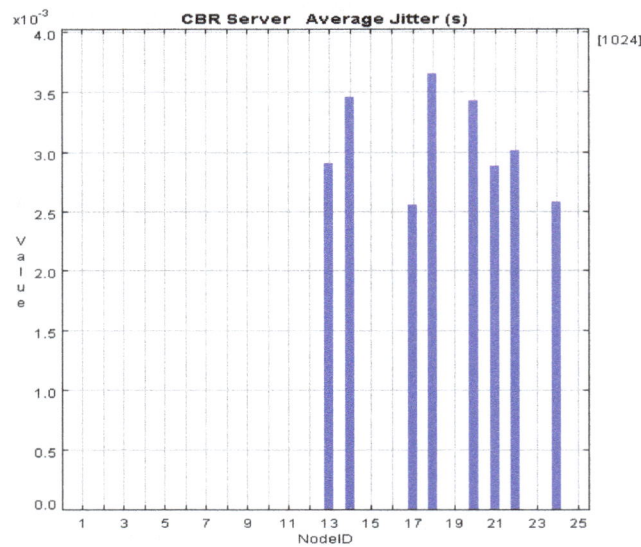

Figure 17. Jitter of Selected Path

The Figure 17 shows packets delay is more in the selected path due to the varying positions of the nodes and more number of hops along the path. Higher jitter can seriously affect the quality of data or streaming audio and/or video.

B. Scenario with 25 Nodes (3 Hops)

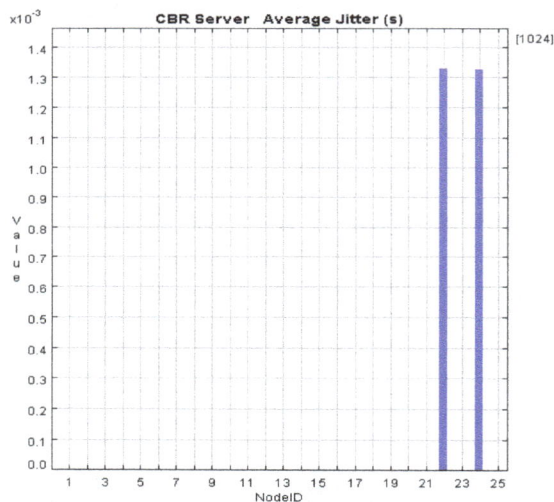

Figure 18. Jitter of Hop Count Optimized Path

When the selected path in Figure 17 is further optimized with hop count feature of developed protocol, jitter is considerably low as shown in Figure 18.

9. CONCLUSION

The paper proposes the development of an energy efficient multipath routing protocol with directional antenna for MANET as an optimization task as well as a multidisciplinary entity encompassing Antenna Engineering, Microwave Link Budget to treat intermediate wireless links, Algorithms to determine the most efficient path for reliable connectivity between the source and destination nodes, comprehensive relation between the RF power and the pertinent Battery Power of the transmit nodes involving conventional electrical power engineering basics as well as choice and optimization of Route metrics to arrive at decision logic for the path selection. This paper presents a comprehensive analysis to link all the multi-disciplinary view points involved in the development of desired multipath routing protocol with requisite technical details. Through the implementation of RF power control, the paper emphasizes and elucidates the importance of Directional antenna in MANET to conserve battery energy and still retaining all other desirable system performance attributes.

Through numerical simulations, the multi-dimensional desirable performance attributes of wireless link such as improved range, improved RSS, reduced RF transmit power and consequent reduced consumption of battery power have been analyzed. Majority of the formulations on Antennas presented in this paper stem from the Friis transmission formula which is well known for its simplicity, importance and empirical nature, that is essential for the design of wireless link Undesirable effect of Beam Pointing Angle error of directional antenna leading to degraded range performance has been illustrated through a typical case study Simulations confirm that significant improvement in energy saving is a reality if the:Range (distance) between the source and the destination nodes is well beyond the maximum directional range of the network operating with its rated peak RF power and Directional Gain is couple of dB higher than Omni Gain. Simulation results on range enhancement with directional antenna reveal that the profile of increased range with variation in antenna gain exhibits an exponential behaviour and can be very effective in the reduction of number of hops Results also confirm that a node with directional antenna requires considerably much lower RF power than that required with an omni directional antenna to retain the same link performance and this in turn leads to smaller battery requirement resulting in energy saving Through typical case studies, the simulation results amply confirm the significant improvement in the life time of the network by using directional antennas.

Simulation results of the implemented multipath Routing protocol associated with built in hop count reduction feature reveal energy saving of 62.11 % for a typical MANET scenario of 25 nodes,5 MB data and data rate of 40 Kbps. From a system perspective, performance of the proposed routing technique can be justified as Highly Energy Efficient Multipath Routing Protocol for MANETs. Contributions that can be derived out of this paper will be of potential utility to address the realistic design challenges and issues pertaining to wireless network in general and energy efficient MANETs in particular. Emphasis on requisite analysis and details on multi-disciplinary aspects of Antennas, Power Computations, algorithm for multi-path determination and Pre Processor to access simulator is a first step in a frame work for development of Network system simulator.

REFERENCES

[1]   Su Yi, Yong Pei and Shivkumar Kalyanaraman,(2003) "On the Capacity Improvement of Ad Hoc Wireless Networks using Directional Antennas", Mobihoc, pp. 108-116.

[2]   R. Ramnathan, (2001) "On the Performance of Ad Hoc Networks with Beamforming Antennas", Mobihoc, pp. 95-105

[3]   A. Nasipuri, K. Li and U. R. Sappidi, (2003) " Power Consumption and Throughput in Mobile Ad Hoc Networks using Directional Antennas", IEEE International Conference on Computer, Communication and Networking

[4]   A. Spyropoulos and C.S. Raghavendra, (2002) "Energy Efficient Communications in Ad Hoc Networks Using Directional Antennas", INFOCOM'02.

[5]   A. Arora, M. Krunz, and A.Muqattash, (2004) "Directional Medium Access Protocol (DMAP) with Power Control for Wireless Ad Hoc Networks", Global Telecommunication Conference

[6]   Amith, K. S., and David, B. (2005) "Routing Improvement Using Directional antennas   in MANETs", IEEE International Conference on Mobile Ad hoc Networks, Fort Lauderdale, Florida, 304-313

[7]   Luz, Y. M., and Vincent Wong W. S. V. (2007) "An Energy-Efficient Multipath Routing Protocol for Wireless Networks",. *International Journal of Communication System* 20(7), pp. 747-766

[8]   Guoqing Li, L. Lily Yang, W. Steven Conner, Bahareh Sadeghi (2003) "Opportunities and Challenges for Mesh Networks Using Directional Antennas".. *Intel Corporation, Communication Technology.*

[9]   Bashir Yahya and Jalel Ben-Othman,(2009) "Robust and Energy Efficient Multipath Routing Protocol for Wireless Networks", Global Telecommunication Conference, pp. 1-7.

[10]   R. Vidhyapriya and P.T. Vanathi, (2006) "Energy Efficient Adaptive Multipath Routing for Wireless Sensor Networks", International Journal of Computer Science, pp. 1-9.

[11]   Chansu Yu, Ben Lee and Hee Yong Youn, (2003) "Energy Efficient Routing Protocols for mobile ad hoc Networks", Wireless Communications and Mobile Computing, pp. 959-973.

[12]   Busola S. Olagbegi and Natarajan Meghanathan, (2010) "A Review of Energy Efficient and Secure Multicast Routing Protocols for Mobile Ad Hoc Networks", International Journal on applications of graph theory in wireless ad hoc networks and sensor networks.

Performance of the IEEE 802.15.4a UWB System using Two Pulse Shaping Techniques in Presence of Single and Double Narrowband Interferences

Rasha S. El-Khamy

Alexandria Higher Institute of Engineering and Technology, Alexandria, Egypt.
E-mail: `rasha_sk@yahoo.com`

ShawkyShaaban, Ibrahim Ghaleb, andHassan Nadir Kheirallah

Dept. of Electrical Engineering,
Faculty of Engineering, Alexandria Univ.,Alexandria 21544, Egypt.

Abstract

In Cognitive radio (CR) applications Ultra-wideband (UWB) impulse radio (IR) signals can be designed such as they can co-exist with licensed primary users. The pulse shape should be adjusted such that the power spectral characteristics not only meet the Federal Communications Commission (FCC) constrains, but also mitigate multiple narrow-band interference at the locations of existing primary users. In this paper, the Parks-McClellan (PM) Algorithm and the Eigen Value Decomposition (EVD) approach for UWB impulse radio waveform shaping are considered. The power spectral density (PSD) and the bit-error-rate (BER) performance of the two methods are compared in the presence of single and double narrow-band interference (NBI). The interference rejection capabilities of the two methods are evaluated and compared for different interference and additive noise levels. In particular, the simulations consider the coexistence of practical IEEE 802.15.4a UWB systems with both IEEE 802.11 wireless LAN systems operating at 5.2 GHz and radio location services operating at 8.5 GHz.

Key words; *Cognitive Radio, UWB, Impulse Radio, Interference Rejection, IEEE 802.15.4a, Parks-McClellanAlgorithm, Eigen Value Decomposition.*

I. Introduction

Ultra-wideband radio is a promising technology for high data rate short-range wireless communication. Compared with the conventional narrowband (NB) communication systems, UWB systems have many advantages, e.g. reduced complexity, low power consumption, immunity to multipath fading, high security, etc. [1]-[3].

Since UWB systems transfer information data by using extremely short duration pulses, they have considerably large bandwidth. FCC regulates UWB systems can exploit the frequencies from 3.1 GHz to 10.6 GHz [4]. From Shannon channel capacity [5], it is evident that UWB systems can achieve higher capacity than any other current wireless communication systems. However, in order to reduce the interference between UWB systems and the existing NB systems, FCC presents a UWB spectral mask to restrict the power spectrum of UWB systems.

The spectrum of a transmitted signal is influenced by the modulation format, the multiple access schemes, and most critically by the spectral shape of the underlying UWB pulse. The choice of the pulse shape is thus a key design decision in UWB systems.

Several pulse design methods of UWB signals have been proposed to let them match with the FCC spectral mask. The simple Gaussian monocycle pulses need to be filtered to meet the FCC spectral mask. This leads that the time duration of the corresponding pulses becomes too long. On the other hand, Gaussian derivatives pulses [6] have fixed features, i.e.; their spectrums are unchangeable once they have been built, making them unable to adjust and adapt their frequency components to avoid frequency colliding. Prolate spherical wave functions [7] and modified Hermite orthogonal polynomials [8] are also used to generate mutually orthogonal pulses that can be used in multiple access schemes. These pulses fit frequency masks with multiple pass-bands. However, they require a high sampling rate that could lead to implementation difficulties.

In this paper, two pulse design methods are discussed and compared as methods of NBI suppression and at the same time overcoming the short-comes of the previously mentioned pulses shapes. The methods give a chance of increasing the UWB transmitted power and enlarging the application range of UWB systems, while meeting the FCC spectral mask. The considered pulse design methods are the Parks-McClellan (PM) Algorithm [9,10], the Eigen Value Decomposition (EVD) approach [11,12].

The paper is organized as follows. Section II presents a filter design method using the PM filter design algorithm. Simulation results of the pulse design for single and double narrowband interference are illustrated. Section III presents the EVD approach for pulse design and shows by simulation how it mitigates multiple NBI. Section IV gives a comparison between the PSD of the EVD approach and the PM pulse design method. Simulations will be done for both single and double NBI. Section V evaluates and compares the bit error rate (BER) performance of both systems in case of single and NBI. Section VI draws the conclusion.

II. Pulse Design Method using the Parks-McClellan (PM) Filter Design Algorithm

A. The PM Algorithm

In this approach the UWB pulses are designed using an adaptive filter design method based on the PM Algorithm. CR Technology[13]is used to suppress the narrowband interference, as well as satisfying the FCC indoor spectral mask.

The PM Algorithm[9,10] was published by James McClellan and Thomas Parks in 1972 as an iterative algorithm for finding the optimal Chebyshevfinite impulse response (FIR) filter. The PM Algorithm is utilized to design and implement efficient and optimal FIR filters. It uses an indirect method for finding the optimal filter coefficients. The goal of the algorithm is to minimize the error in the pass and stop bands by utilizing the Chebyshev approximation. The PM Algorithm is a variation of the Remez Algorithm[9], with the change that it is specifically designed for FIR filters and has become a standard method for FIR filter design.

The use of PM method to design UWB pulses with single or double notches at the narrow band interference frequency was considered in [14]. The spectrum of the adaptive pulse, $S(f)$, can be expressed as:

$$S(f) = \bar{H}(f) - N_n(f) \tag{1}$$

Where $\bar{H}(f)$ and $N_n(f)$ are the spectra of the UWB needed to satisfy the FCC mask and the narrowband interference spectrum to be cancelled. The estimated optimal pulse spectrum, $\tilde{S}(f)$ can be expressed as a polynomial of order R:

$$\tilde{S}(f) = \sum_{i=0}^{R} a_i f^i \tag{2}$$

The observational error is defined as:

$$e(f) = \gamma(f)[S(f) - \tilde{S}(f)] \tag{3}$$

where, $\gamma(f)$, is a suitable weighting function. The remez(..) function in MATLAB can then be used to get the solution that minimizes the error.

B. Simulation Results of Pulse Design:

*iSingle Narrowband Interference:*The adaptive pulse, the normalized spectrum and the resultant designed spectrum are shown in Fig. 1. It was generated by setting the remez(N,F,M,W) function with N=90 sample points. The frequency vector F, its corresponding amplitude value M, and the weight vector W are as follows:
F1=[0,0.21,0.2583,0.39,0.432,0.46, 0.48,0.73,0.76,1];
M1= [0,0,1,1,0,1,1,1,0.316,0.316]; W1=[6,1,6,1,3].

The NB interference is assumed at 5.2 GHz. It is clear from Fig. 1 that the spectrum fit the normalized FCC mask while the narrowband interference at *5.2 GHz* is suppressed.

*ii. Double Narrow-band Interference:*Now we assume *two* NBI signals interfering with the UWB band centered at frequencies *5.2* and *8.5* GHz. Using the same pre-described remez function, the generated pulse in this case and its PSD are obtained as shown in Fig. 2.

III. Eigen-Value Decomposition (EVD) Approach

Another pulse design method based on Eigen-value decomposition [11,12]is considered in this paper as another method to suppress single or multiple narrow-band interferences located in the UWB band.

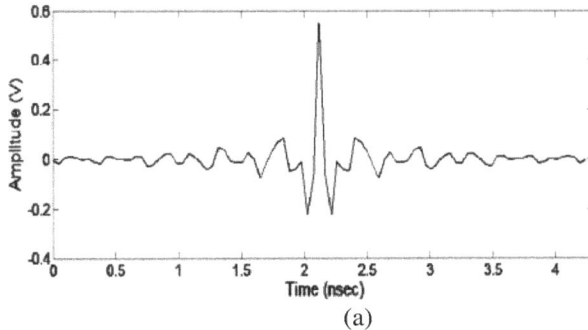

(a)

(b)

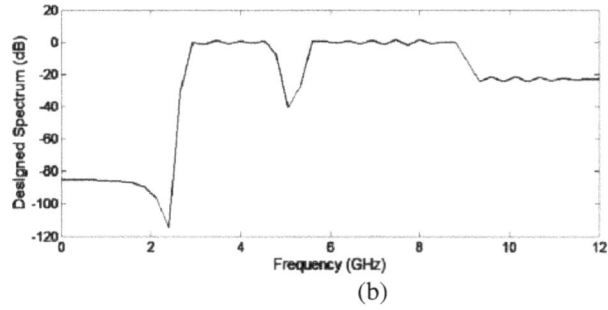

Fig. 1. Adaptive Generated Pulse for Single NBI Rejection using the PM Algorithm
a) Pulse shape b) Normalized PSD of generated pulse.

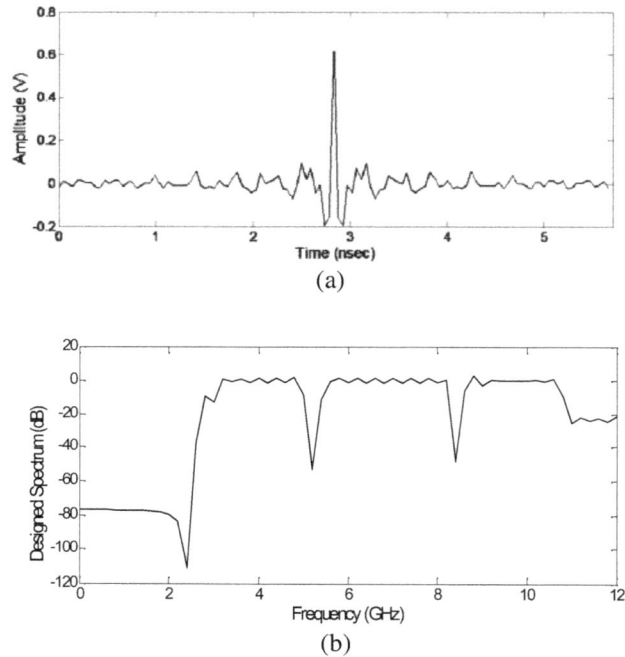

(a)

(b)

Fig. 2 Adaptive Generated Pulse for Multiple NBI Rejection using the PM Algorithm:
a) Normalized Pulse shape   b) PSD of generated pulse

The time response s(t) of the ideal pulse covering the frequency band $f_L \leq f \leq f_H$, where: f_L=3.1GHz, f_H=10.6GHz is obtained by the inverse Fourier transform of the FCC ideal mask $H(f)$ as,

$$h(t) = 2f_H \text{sinc}(2f_H t) - 2f_L \text{sinc}(2f_L t) \tag{4}$$

where, sinc(x)=sin(x)/ x. The UWB pulses $s(t)$ can thus be generated by convolution:

$$\lambda s(t) = \int_{-\infty}^{\infty} s(\tau)h(t-\tau)d\tau \tag{5}$$

where, λ is a constant (Eigen value). By sampling at a rate of N samples per pulse period T_m, , equation (5) can be expressed as an Eigen Value Problem, namely,

$$\lambda . \boldsymbol{s} = \boldsymbol{H}. \boldsymbol{s} \tag{6}$$

where, the vector s represents the samples of discretized UWB pulse, H is a Hermitian matrix constructed by the discrete samples of h(t).

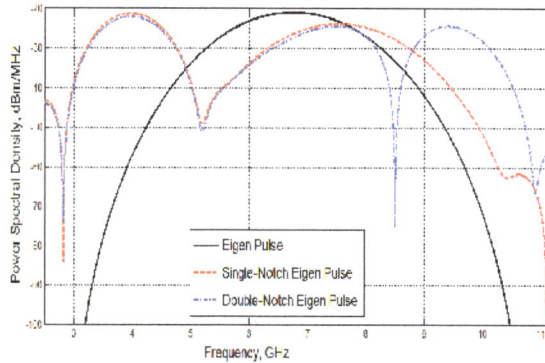

Fig. 3. PSD of the Eigen pulse in absence of an Interferer, and compared to that of the Designed Single and Double Notch Eigen Pulse

The extension of the designed Eigen pulses to affect nulls at the frequencies of narrowband interference was discussed in [15]. Comparison between the obtained spectra of the ideal pulse, the pulse needed to mitigate single narrowband interference of the IEEE 802.11 wireless LAN systems operating at 5.2 GHzand the pulse needed to mitigate double narrow band interferences of the radio location services operating at 8.5 GHz are shown in Fig. 3 [15].

IV. PSD Comparison between (EVD) Method and (PM) Filter Design Algorithm

A- Single Narrowband Interference:

The PSD of Single Notch Modified Eigen Pulse is compared to the PSD of a Single Notch adaptive PM Pulse in Fig. 4. The NB interference is assumed at *5.2 GHz*.

The used design parameters for the generation of the Single Notch Eigen Pulse are: sampling frequency f_s=120 GHz, f_L=3.1GHz, f_M=4.9GHz, f_N=4 GHz, and f_H=11.1 GHz.

The PSD of adaptive PM pulse is generated as seen in Fig. 4 by setting the remez (N,F,M,W) function with N=90 sample points, sampling frequency of 30 GHz, frequency vector F, its corresponding amplitude value M, and the weight vector W are as follows:
F1= [0   0.168   0.20664   0.33   0.35   0.37   0.384   0.68667   0.70667   1] ;
M=[0   0   1   1   0   1   1   1   0   0]

B- Double Narrowband Interference:

The PSD of Double Notch Eigen Pulse iscompared to the PSD of a Double Notch adaptive PM Pulse as shown in Fig. 5. The NB interference is assumed at *5.2 GHz* and *8.5 GHz*.

The used design parameters for the generation of the Double Notch Eigen Pulse are: f_s=120GHz, f_{L1}=3.1GHz, f_{H1}=4.9GHz, f_{L2}=4GHz, f_{H2}=11.1 GHz, f_{L3}= 7.9 GHz, f_{H3}=10.7 GHz.

The PSD of adaptive PM pulse is generated as seen in Fig.5 by setting the remez (N,F,M,W) function with N=90 sample points, sampling frequency of 30 GHz, frequency vector F, its corresponding amplitude value M, and the weight vector W are as follows:

F1= [0   0.168   0.20664   0.33   0.35   0.368   0.384   0.55   0.56667
0.59   0.6   0.68667   0.70667]
M = [0   0   1   1   0   1   1   1   0 0]

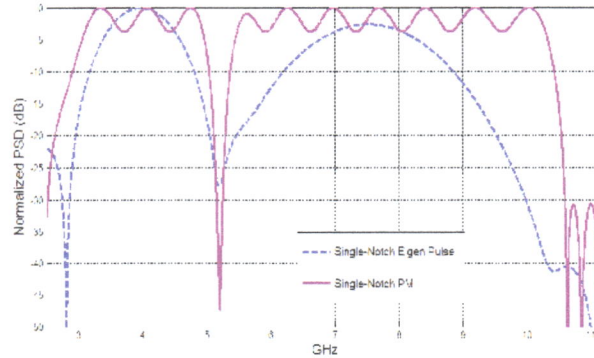

Fig. 4.PSD of Single Notch Adaptive PM Pulse, and compared to PSD of Single Notch EigenPulse

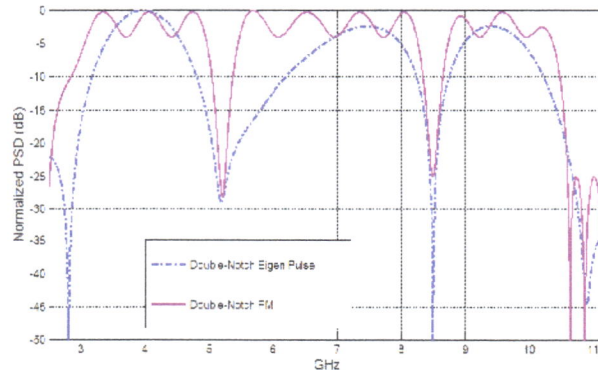

Fig. 5.PSD of Double Notch Adaptive PM Pulse, and compared to PSD of Double Notch Eigen Pulse

V.Bit Error Rate (BER) Performance Comparison between(EVD) Method and (PM) Filter Design Algorithm

The BER performance of the IEEE 802.15.4a UWB system transmitting on IEEE 802.15.4a channel is evaluated. The IEEE 802.15.4a UWB channel model is deployed[16].The IEEE 802.15.4a UWB channel model is deployed. The UWB channel model is a Saleh-Valenzuela (SV) model modified to take into account different measurement and simulation parameters [23]. Our simulations in this paper assume the residential LOS model CM1, with parameters as given in Table I of [23].

The IEEE 802.15.4a transmitter is specified by the standard and is explained in details in [16-23]. The receiver deployed is a coherent detector assuming perfect UWB channel estimation available at the receiver [21].

A- Single Narrowband Interference:

We consider the coexistence of practical IEEE 802.15.4a UWB systems with IEEE 802.11 wireless LAN systems operating at 5.2 GHz.

In Fig. 6, BER performance is evaluated for each of the ideal Eigen Pulse (in absence of interference), Single Notch Eigen Pulse and the Single Notch adaptive PM Pulse. The BER performance is compared versus the SIR, and at Signal to Noise Ratio (SNR) fixed at E_b/N_0 of 9 dB.It is clear from the figure that the Single Notch adaptive PM Pulse outperforms the Eigen-Value pulse whether there is a single- notch interference or in its absence.

In Fig. 7, BER performance evaluation is repeated. This time, the BER performance is simulated versus the SNR and at Signal to Interference Ratio (SIR) fixed at 5 dB. The same results appear here also; i.e: the BER of the Single Notch adaptive PM Pulse is the least among the three methods at any SNR value

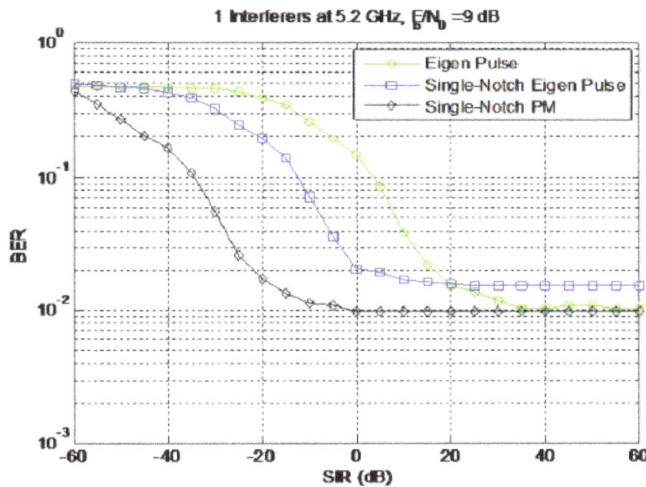

Fig.6.BER vs SIR at E_b/N_0 of 9 dB in the presense of a single narrow band interference

B- Double Narrowband Interference:

Here, the coexistence of practical IEEE 802.15.4a UWB systems is considered with both IEEE 802.11 wireless LAN systems operating at 5.2 GHz and radio location services operating at 8.5 GHz.

Simulations are repeated in Fig. 8 but comparisons are extended to include Double NBI at 8.5 GHz. BER performance is evaluated for each of the ideal Eigen Pulse (in absence of interference), Double Notch Eigen Pulse and the Double Notch adaptive PM Pulse. The BER performance is compared versus the SIR. The Signal to Noise Ratio (SNR) is fixed at E_b/N_0 of 3dB. Results show that the Double Notch Eigen Pulse outperforms the Double Notch adaptive PM Pulse at any SIR value.

In Fig. 9, BER performance evaluation is repeated. This time, the BER performance is simulated versus the SNR. Signal to Interference Ratio (SIR) is fixed at 5 dB . Results show that the performance of Double Notch Eigen Pulse is nearly the same as that of Double Notch adaptive

PM Pulse; BER is nearly the same at any SNR value. However, if the SNR is deeply lowered to -20 dB, results show that the Double Notch Eigen Pulse gives the better performance.

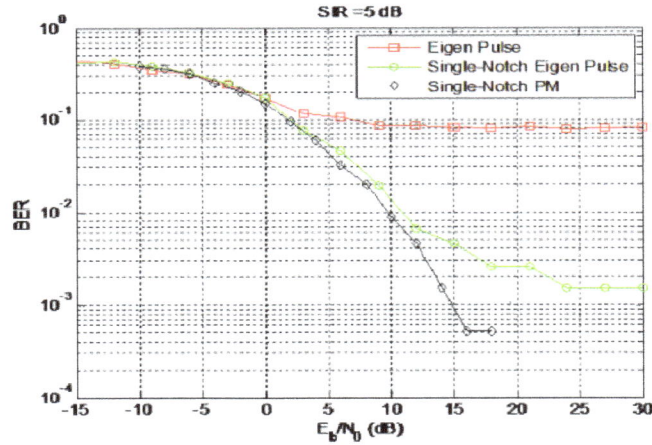

Fig.7. BER vs SNR at fixed SIR of 5 dB

VI. Conclusion

In this paper two methods are presented to suppress NBI located in the UWB band without the need of lowering the UWB pulse PSD. These methods not only give a solution to the co-existence between UWB systems and existing narrowband systems, but also give a chance of increasing the UWB transmitted power and enlarging the application range of UWB systems, while meeting the FCC spectral mask. The PSD and BER comparisons between both the EVD approach and the PM pulse design algorithm for single and double NBI are evaluated. Results showed that for single NBI, the PM algorithm outperforms the EVD method. While in the case of double NBI, the performance of both methods are nearly the same, except at very low values; SNR= -20dB, performance of the PM algorithm degrades alot.

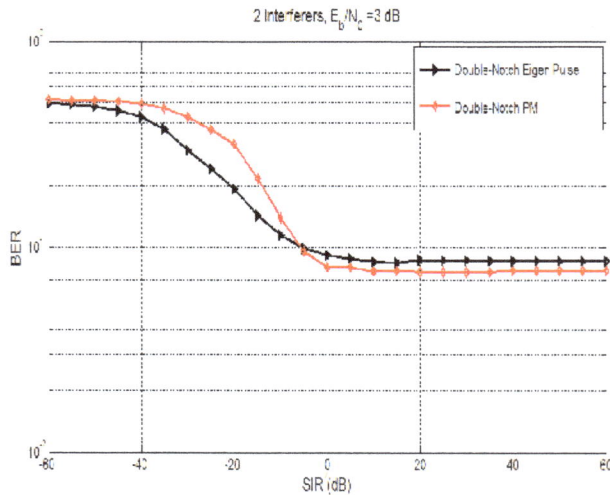

Fig.8.BER vs SIR at atE_b/N_0 of 3 dB

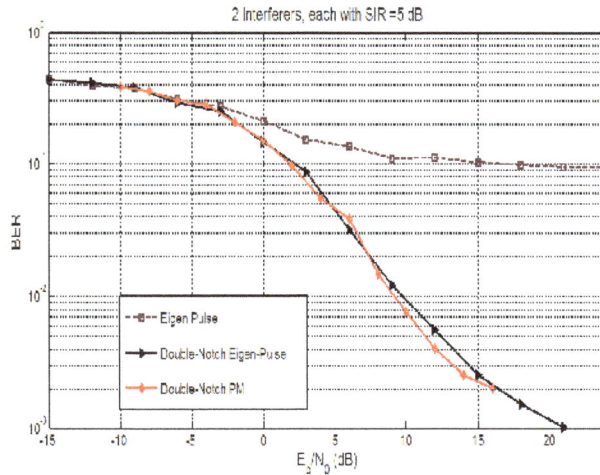

Fig.9.BER vs SNR at fixed SIR of 5 dB

REFERENCES

[1] *Intel White Paper*, "Ultra-wideband (UWB) Technology: Enabling high-speed wireless personal area networks," 2004.Online].Available:http://www.intel.com/technology/comms/uwb

[2] C. Chong, F. Watanabe, and H.Inamura , "Potential of UWB Technology for the Next Generation Wireless Communications *(Invited Paper)"*, *IEEE Ninth International Symposium on Spread Spectrum Techniques and Applications*, pp.442-49, 2006.

[3] H. Arslan. Chen, and M. Di Benedetto, *"Ultra Wideband Wireless Communication"*, John Wiley & Sons, Inc. , 2006.

[4] "Revision of Part 15 the Commission's rules regarding ultra-wideband transmission systems", FCC, ET Docket, pp.98-153, 2002.

[5] T. M. Cover and j. a. Thomas, "Elements of Information Theory", NewYork: Wiley, 1991.

[6] T. Phan , V. Krizhanovskii , S. Han , and S. Lee , and N. Kim, "7th Pulse Derivative Gaussian Pulse Generator for Impulse Radio UWB", *Auto-ID Labs White Paper*.

[7] Y. Lu, H. B. Zhu, "UWB pulse design method based on approximate prolate spheroidal wave function", *Journal of Communications*, Vol.26, pp. 60-64, 2005.

[8] M. Ghavami, L. B. Michael and R. Kohno, "Hermite function based orthogonal pulses for UWB communication", Proc. Int. Symp. on Wireless Personal Multimedia Communications, pp. 437–440, Aalborg, September 2001.

[9] Xiaolin Shi, "Adaptive UWB Pulse Design Method for Multiple Narrowband Interference Suppression", 2010 IEEE International Conference on Intelligent Computing and Intelligent Systems (ICIS), 29-31 Oct. 2010.

[10] X. Luo, L. Yang, and G. B. Giannakis, "Designing optimal pulse-shapers for ultrawideband radios," Journal of Communications and Networks, vol. 5, no. 4, pp. 344–353, December 2003.

[11] B. Das, S. Das, " Interference Cancellation Schemes in UWB Systems used Wireless Personal Area Network based on Wavelet based Pulse Spectral shaping and Transmitted Reference UWB using AWGN Channel Model", International Journal of Computer Applications, Vol. 2 , No.2, pp. 0975 – 8887, May 2010.

[12] A. Taha., K. M. Chugg, "A theoretical study on the effects of interference UWB multiple access impulse radio Signals," Systems and Computers 2002 Conference Record of the Thirty-Sixth Asilomar Conference., Vol.1, pp.728-732, Nov 2002.

[13] V. D. Chakravarthy, A. K. Shaw, and M. A. Temole, "Cognitive radio- An adaptive waveform with spectral sharing capability", IEEE Wireless Communications and Networking Conference, pp.724-729, 2005.

[14] R.S. El-Khamy, S.E. Shaaban, H.N. Kheirallah, and I.A. Ghaleb, " UWB Impulse Radio Waveform Shaping Techniques for Narrow-Band Interference Rejection", *International Journal of Engineering and Innovative Technology*, Vol 2, Issue 9, pp.134-140 ,March 2003.

[15] R. S. El-Khamy, S. E. Shaaban, H. N. Kheirallah, and I. A. Ghaleb, "An Interference-Aware Cognitive Radio UWB(IA-CR-UWB) System Using Eigen Pulse Shaping", *The International Conference on Technological Advances in Electrical, Electronics and Computer Engineering (TAEECE2013)*,Mevlana University, Turkey, pp. 17-22, May 2013.

[16] M. Flury, R. Merz, J.-Y. Le Boudec, and J. Zory, "Performance evaluation of an IEEE 802.15.4a physical layer with energy detection and multi-user interference", *Proc. IEEE Intl. Conf. Ultra-Wideband (ICUWB)*, pp.663-668,Singapore, Sept. 2007.

[17]M. Ghavami, L. B. Michael, R. Kohno, "Ultra Wideband Signals and Systems in CommunicationEngineering", 2nd Edition, Wiley, Mar. 2007.

[18] IEEE Computer Society, Amendment to IEEE Std. 802.15.4, IEEE Std. 802.15.4a-2007, Aug. 31, 2007.

[19] Z. Ahmadian, and L. Lampe, "Performance Analysis of the IEEE 802.15.4a UWB System",*IEEE Transactions onCommunications*, Vol. 57, No. 5, pp. 1474-1485 May 2009.

[20] V. Lottici, A. D'Andrea, and U. Mengali, "Channel estimation for ultra-wideband communications," *IEEE Journal on Selected Areas in Communications*, vol. 20, no. 9, pp. 1638–1645, 2002.

[21] Z. Tian and B. Sadler, "Weighted energy detection of ultra-wideband signals" ,*IEEE SPAWC 05*, June 2005, pp. 1068–1072.

[22] IEEE 802 Working Group, "Standard for Part 15.4: Wireless Medium Access Control (MAC) and Physical Layer (PHY) Specifications for Low Rate Wireless Personal Area Networks (LR-WPANs)," *IEEE Std 802.15.4a, 2007.*

[23] A.F. Molisch, D. Cassioli, C. Chong, S. Emami, A. Fort, B. Kannan, J. Karedal, J. Kunisch, H. Schantz, K. Siwiak and M. Win, "A Comprehensive Standardized Model for Ultrawideband Propagation Channels," *IEEE Transactions on Antennas and Propagation,* vol. 54, no. 11, pp. 3151-3166, 2006.

11

A New UWB System Based on a Frequency Domain Transformation Of The Received Signal

Karima Ben Hamida El Abri and Ammar Bouallegue

Syscoms Laboratory, National Engineering School of Tunis, Tunisia
Emails: enitkarima@yahoo.fr, ammar.bouallegue@enit.rnu.tn

Abstract

Differential system for ultra wide band (UWB) transmission is a very attractive solution from a practical point of view. In this paper, we present a new direct sequence (DS) UWB system based on the conversion of the received signal from time domain to frequency domain that's why we called FDR receiver. Simulation results show that the prposed receiver structure outperforms the classical differential one for both low and high data rate systems.

1 Introduction

UWB technology has been proposed as a viable solution for high speed indoor short range wireless communications systems due to multipath channel. Moreover, UWB technology can offer simultaneously high data rate and low power implementation. In order to meet the spectrum mask released by the Federal Communications Commission (FCC) and to obtain the adequate signal energy for reliable detection, each information symbol is represented by a train of very short pulses, called monocycles. Each one is located in its own frame.

Regarding demodulation, two types of detectors are considered for UWB systems: coherent and non coherent receiver. The former needs channel information estimation, which is not always achievable, especially at low SNR, even if all the transmitted energy is used for that [9]. In these cases, non-coherent detection is more suitable. That is why, much researches on UWB has focused on non-coherent systems such as: Transmitted Reference (TR) and Differential system. In TR UWB system [10], the transmitted signal consists of a train of pulses pairs. Over each frame, the first pulse is modulated by data. The second one is a reference pulse used for signal detection at the receiver. Reception is made by delaying the received signal and correlating it with the original version. The simplicity of this receiver is very attractive. Nevertheless, TR systems waste half of the energy to transmit reference signals. That's why the TR systems are replaced by differential systems, where detection is achieved by correlating the received signal and its replica delayed by a period D (D can be the symbol period [11], the frame period [12] or a function of chip, symbol and frame period [12] [13]).

In this paper, we introduce a new type of receiver that we called FDR receiver. The structure is very simple. It is based on the projection of the received signal in a basis of functions in order to transform the input of the receiver from time domain to frequency domain, followed by multiplication with a spreading code. With a judicious choice of the basis and the code used, we show that we have not only transform the domain of the received signal. But, we have also transform in someway the PAM (Pulse Amplitude Modulation) modulation used in the transmitter to a PPM (Pulse Position Modulation) modulation. In fact, the useful energy is no longer concentrated in the begining of the frame but it depends on the transmitted data. And this behaviour is similar to PPM modulation.

The remainder of the paper is organized as follows. In section 2, we begin by an overview of the UWB channel model. Then, in section 3, we present the conventionnal differential DS-UWB system followed by description of the proposed receiver in section 4. As for simulation results and comparisons, they are presented in section 5. Finally, a conclusion is given in section 6.

2 UWB channel model

The basic conditions of UWB systems differ according to applications. It is based on the conventionnal Saleh and Valenzuela (S-V) channel model [15]. We distinguish two kind of propagation environments: outdoor and indoor propagation. The former is dominated by a direct path while the latter is made of a dense multipath. In this work, we consider the IEEE 802.15 UWB indoor channel [8], where multipath arrivals are grouped into two categories:cluster arrivals, and ray arrivals within each cluster.

$$h(t) = \sum_{l=0}^{L-1} \sum_{k=0}^{K-1} \alpha_{k,l} \delta(t - T_l - \tau_{k,l}) \qquad (1)$$

where:

- $\alpha_{k,l}$ denotes the multipath gain coefficient.
- T_l is the lth cluster arrival time.
- $\tau_{k,l}$ represents the delay of kth multipath component inside the cluster l.
- $\delta(t)$ is the Dirac delta function.

The UWB channel given in $(eq.)$ can be modeled as a tapped delay line defined as follows:

$$h(t) = \sum_{l=0}^{L-1} \alpha_l \delta(t - \tau_l) \qquad (2)$$

with:

- α_l denotes attenuation of each path.
- τ_l represents the delay of kth path. It satisfies $\tau_0 < \tau_1 < \cdots < \tau_L$.

3 Related work: Differential DS-UWB system

3.1 Modulation

In the UWB transmission, every symbol is transmitted by employing N_f short pulses $\omega_T(t)$, each with an ultra short duration T_ω of the order of nanosecond and normalized energy. The pulses are transmitted once per frame.

We propose to use a DS-UWB system which is based on a train of short pulses multiplied by a

spreading sequence [8]. Each pulse is located in the begining of the frame.
The transmitted signal is given by:

$$s(t) = \sum_j d_j C_{\left\lfloor \frac{j}{N_f} \right\rfloor} \omega_T(t - jT_f) \qquad (3)$$

where:

- d_j is the differentially encoded bit given by: $d_j = d_{j-1}.b_{\left\lfloor \frac{j}{N_f} \right\rfloor}$, where $b_{\left\lfloor \frac{j}{N_f} \right\rfloor}$

represents the random binary data symbol sequence in frame j taking values ± 1, with equal probability.
- $\{C\}$ is the spreading sequence of length N_f. Each $C_{\left\lfloor \frac{j}{N_f} \right\rfloor}$ is used to code the pulse

contained in the jth frame and it takes values ± 1.
- T_f is the frame duration verifying $T_f \gg T_\omega$.

After propagation in the channel, the received signal $r(t)$ is given by:

$$\begin{aligned} r(t) &= s(t) \otimes h(t) + n(t) \\ &= \sum_j d_j C_{\left\lfloor \frac{j}{N_f} \right\rfloor} \omega_R(t - jT_f) + n(t) \end{aligned} \qquad (4)$$

Where:

- $n(t)$, is an additive white gaussian noise (AWGN) with two side spectral density $\frac{N_0}{2}$

and zero mean.
- $\omega_R(t)$ represents the received waveform of each symbol defined as: $\omega_R(t) = \sum_{l=0}^{L-1} \alpha_l \omega_T(t - \tau_l)$

3.2 Demodulation

To detect the emitted symbols b_m, we suggest to use a differential receiver based on the correlation of the received signal given in eq.4 with its replica delayed by a frame duration T_f.
A block diagram of the receiver is presented in Fig.1 .

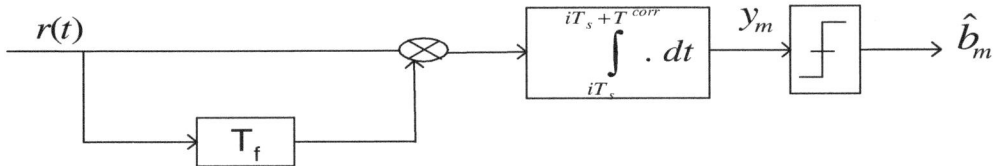

Figure 1: Differential Receiver structure.

The output y_m of the frame differential receiver for the mth bit is given by:

$$\begin{aligned}y_m &= \int_{mT_s}^{(m+1)T_s} r(t).r(t-T_f)dt \\ &= \sum_{j=0}^{N_f-1} \int_{jT_f}^{jT_f+T_{corr}} r(t).r(t-T_f)dt\end{aligned} \tag{5}$$

where:
- T_{corr}: is the integration window, $T_{corr} = T_\omega + T_{mds}$ (T_{mds} is the maximum delay spread of the channel).

Let's pose:

$$\begin{aligned}y_j &= \int_{jT_f}^{jT_f+T_{corr}} r(t).r(t-T_f)dt \\ &= s(j) + \sum_{i=1}^{3} n_i(j)\end{aligned} \tag{6}$$

In eq. 6, $s(j)$ is the desired signal in frame j and $\{n_i(j)\}_{i=1:3}$ are noise terms due to signal-noise correlation and noise cross noise correlation.

The expression of the desired signal s(j) is given by:

$$s(j) = b_{\left\lfloor \frac{j}{N_f} \right\rfloor} c_{\left\lfloor \frac{j}{N_f} \right\rfloor} c_{\left\lfloor \frac{j-1}{N_f} \right\rfloor} R_\omega(0) \tag{7}$$

Where $R_\omega(0) = \int_0^{T_{corr}} \omega_T^2(t)dt$.

As for noise terms, they are listed below:

- $n_j(1) = d_j c_{\left\lfloor \frac{j}{N_f} \right\rfloor} \int_0^{T_{corr}} \omega_T(t)n(t+(j-1)T_f)dt$

- $n_j(2) = d_{j-1} c_{\left\lfloor \frac{j-1}{N_f} \right\rfloor} \int_0^{T_{corr}} \omega_T(t)n(t+jT_f)dt$

- $n_j(3) = \int_{jT_f}^{jT_f+T_{corr}} n(t)n(t-T_f)dt$

To find the estimation of the mth bit, we just have to multiply y_j before summation by $c_{\left\lfloor \frac{j}{N_f} \right\rfloor} c_{\left\lfloor \frac{j-1}{N_f} \right\rfloor}$. Then, we only have to take the sign of $y_m \Rightarrow \hat{b}_m = sign(y_m)$

4 The proposed system based on a FDR receiver

4.1 Modulation

The transmitter structure is the same as the structure described in section 3.
We can write the expression of the emitted signal s(t) as follows:

$$s(t) = \sum_j d_j \omega_T(t-jT_f) \tag{8}$$

Where j is the index of the jth frame, $j \in [0, N * N_f)$ and N represents the total number of emitted bits.

Let's pose: $d_j = c_{\left\lfloor \frac{j}{N_f} \right\rfloor} \cdot b_{\left\lfloor \frac{j}{N_f} \right\rfloor} \cdot d_{j-1}$

For simplicity, we pose: $m = \left\lfloor \frac{j}{N_f} \right\rfloor$ which represents the mth emitted bit, $m \in [0, N)$ and $i = j - mN_f \in [0, N_f)$

And so, the design of the jth code can be obtained as described in Fig.2 .

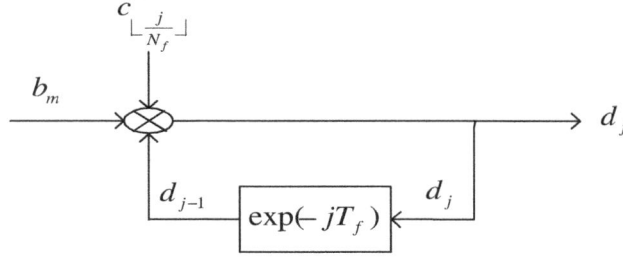

Figure 2: Description of code construction.

By induction, we obtain :

$$d_j = c_i' \cdot b_m^{i+1} \cdot d_{mN_f-1}$$

where $c_i' = \left[c_0', c_1', \ldots, c_{N_f-1}' \right] = \prod_{l=0}^{i} c_l$

Now, we will take the upper value of i to express d_j as a function of d_{-1}. In this case $(i = N_f - 1)$, we obtain:

$$\begin{aligned} d_j &= d_{mN_f+i} \\ &= d_{(m+1)N_f-1} \\ &= c_{N_f-1}' \cdot b_m^{N_f} \cdot d_{mN_f-1} \end{aligned}$$

$\Rightarrow$

$$\begin{aligned} d_{mN_f-1} &= c_{N_f-1}' \cdot b_{m-1}^{N_f} \cdot d_{(m-1)N_f-1} \\ &= (c_{N_f-1}')^2 \cdot b_{m-1}^{N_f} \cdot b_{m-2}^{N_f} \cdot d_{(m-2)N_f-1} \end{aligned}$$

By induction, we find the following relation:

$$d_{mN_f-1} = (c_{N_f-1}')^m \cdot (\prod_{l=0}^{m-1} b_l)^{N_f} \cdot d_{-1}$$

And so, the expression of d_j becomes:

$$d_j = c_i'.b_m^{i+1}.d_{mN_f-1} = c_i'.b_m^i.a_m$$

Where:

$$a_m = (c_{N_f-1}')^m.(\textstyle\prod_{l=0}^{m-1} b_l)^{N_f}.d_{-1}.b_m \tag{9}$$

Hence, the expression of s(t) can be rewritten as follows:

$$\begin{aligned} s(t) &= \textstyle\sum_j d_j\omega(t - jT_f) \\ &= \textstyle\sum_j a_m.c_i'.b_m^i.\omega(t - jT_f) \end{aligned} \tag{10}$$

4.2 Demodulation

After propagation in the UWB channel, the transmitted signal arrives at the receiver in distorted waveforms and so, the received signal r(t) is given by:

$$\begin{aligned} r(t) &= s(t) \otimes h(t) + n(t) \\ &= \textstyle\sum_j a_m\, c_i'\, b_m^i \omega_R(t - jT_f) + n(t) \end{aligned} \tag{11}$$

Where:

- $\otimes$ represents the convolution operator.
- $h(t)$ is the channel impulse response.
- $n(t)$ is an Additive White Gaussian Noise, with power density $\frac{N_0}{2}$.
- $\omega_R(t)$ is the received waveform after propagation in the channel,
  $$\omega_R(t) = \omega_T(t) \otimes h(t).$$
- m and i are defined for simplification. They are given by: $m = \left\lfloor \frac{j}{N_f} \right\rfloor$ and $i = j -$

mN_f

The receiver model is based on the projection of the received signal r(t) in the frequency domain. The receiver structure is described in Fig.3 .

At the input of the receiver, the signal $r_m(t)$ which represents the received signal in the mth bit, is composed of N_f signal $r_m(l, t)$ which corresponds to the signal in the frame l.

$$\begin{aligned} r_m(t) &= \textstyle\sum_{l=0}^{N_f-1} r_m(l, t) \\ &= \textstyle\sum_{l=0}^{N_f-1} r_m(t + l.T_f) \end{aligned} \tag{12}$$

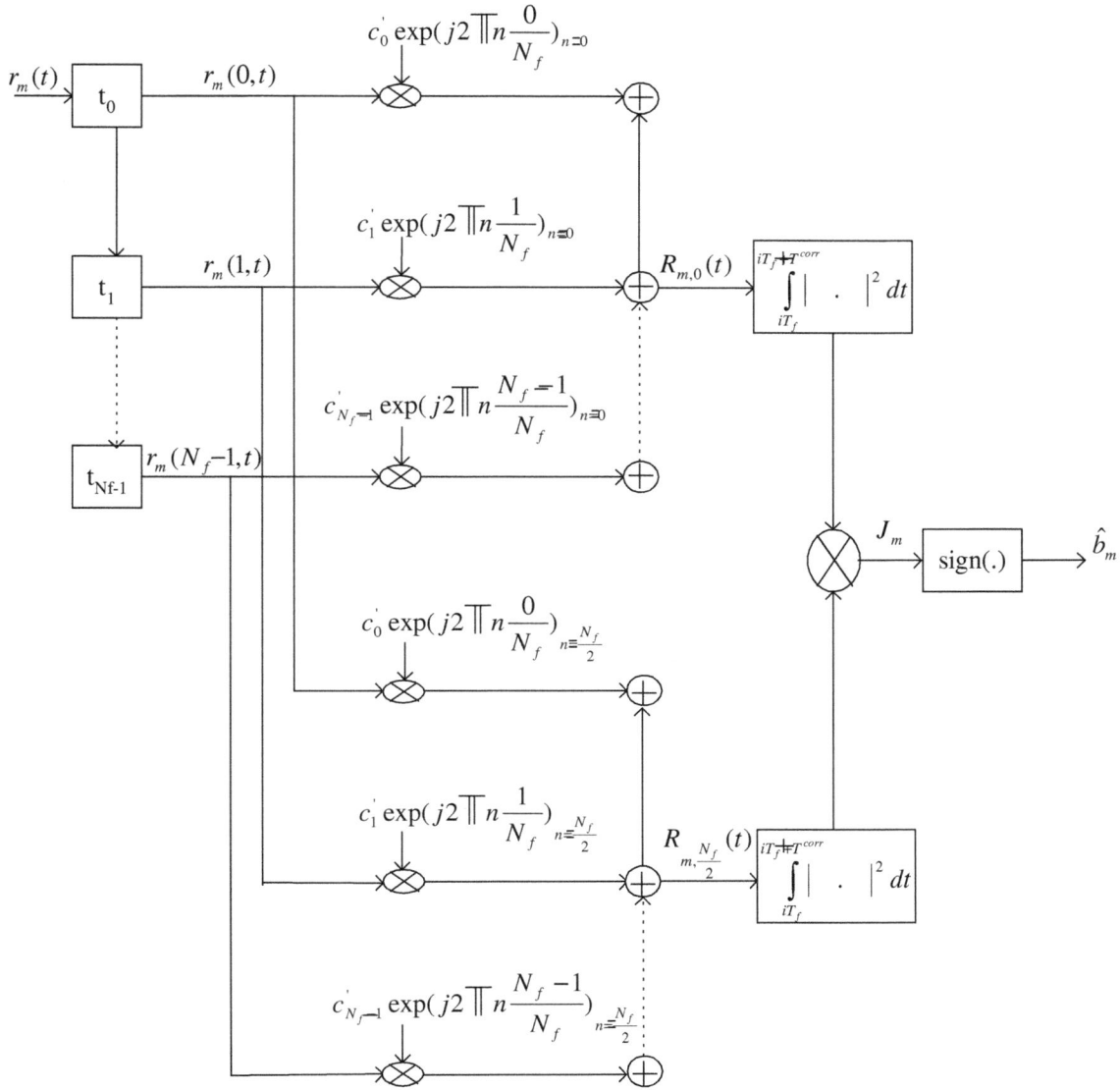

Figure 3: The FDR receiver structure.

Each portion of $r_m(t)$ can be written as:

$$r_m(l,t) \quad = s_m(l,t) + n_m(l,t)$$

Where $s_m(l,t)$ is the desired signal in the mth bit, in the frame j. It is given by:

$$s_m(l,t) \quad = s_m(t + l.T_f)$$
$$= a_m.c_i'.b_m^i.\omega_R(t)$$

Then, we project each portion of $r_m(t)$ in the new basis and multiply it by the code c_l'.

Hence, we obtain the following signal for bit m:

$$
\begin{aligned}
R_{m,n}(t) &= \sum_{l=0}^{N_f-1} c_l' \, r(l,t) \, exp(j2\pi \frac{nl}{N_f}) \\
&= S_{m,n}(t) + N_{m,n}(t)
\end{aligned}
\tag{13}
$$

Where:

- $n \in [0, N_f)$
- $S_{m,n}(t)$ and $N_{m,n}(t)$ are respective projections of $s(l,t)$ and $n(l,t)$ in the new

basis.

We show in the Appendix that the estimation of bit m ($\hat{b}_m$) is specified using only two values of $R_{m,n}(t)$ which are: $R_{m,0}(t)$ and $R_{m,\frac{N_f}{2}}(t)$. In fact, the useful energy is concentrated in $R_{m,0}(t)$ if the transmitted bit $b_m = 1$ or in $R_{m,\frac{N_f}{2}}(t)$ if the transmitted bit $b_m = -1$. As for noise component, they are located in the other cases. We have:

- $b_m = 1 \Rightarrow$

$$
|R_{m,n}(t)|^2 = N_f^2 \, \omega_R^2(t) + a_m N_f \omega_R(t)(N_{m,n}(t) + N_{m,n}(t)^*) + |N_{m,n}(t)|^2, if \ n = 0
$$
$$
|R_{m,n}(t)|^2 = |N_{m,n}(t)|^2, if \ n \neq 0
$$

- $b_m = -1 \Rightarrow$

$$
|R_{m,n}(t)|^2 = N_f^2 \, \omega_R^2(t) + a_m N_f \omega_R(t)(N_{m,n}(t) + N_{m,n}(t)^*) + |N_{m,n}(t)|^2, if \ n = \frac{N_f}{2}
$$
$$
|R_{m,n}(t)|^2 = |N_{m,n}(t)|^2, if \ n \neq \frac{N_f}{2}
$$

This behaviour is similar to a PPM modulation using frames of duration $N_f T_f$ and pulses having energy N_f times more than the transmitted pulses. And so, the detection is based on an energy collector intergator.

Hence, we transform a PAM modulation to a PPM modulation. This explain the performance of the proposed system using the FDR receiver which outperforms the differential receiver. Therefore, to decide of the value of $\hat{b}_m$, we have to use the following criterion:

$$
\begin{aligned}
J_m &= \int_0^{T^{corr}} |R_{m,0}(t)|^2 dt - \int_0^{T^{corr}} \left| R_{m,\frac{N_f}{2}}(t) \right|^2 dt \\
&= \int_0^{T^{corr}} \left[|R_{m,0}(t)|^2 - \left| R_{m,\frac{N_f}{2}}(t) \right|^2 \right] dt
\end{aligned}
\tag{14}
$$

The decision is given by the sign of the criterion J_m.

5 Simulation

In this section, we present the parameters used in the simulation of the two systems implemented: the DS-UWB system using the differential receiver and the DS-UWB system using the proposed FDR receiver. Results are given in terms of Bit Error Rate (BER) as a function of Signal to Noise Ratio (SNR).

The simulations are performed through numerical Monte Carlo simulations. In each trial, the following suppositions are made:

• The monocycle $\omega_T(t)$ is normalized so that the total symbol energy is unity, with duration $T_\omega = 0.8ns$.

• The spreading codes are generated randomly from ± 1, with equal probability.

• We simulate the multipath channel using the model CM1 from [8]. The channel is assumed to be time invariant within a burst of symbols. The maximum delay spread of the channel is $5ns$. To avoid inter frame interference, we truncated the channel to $\frac{T_f}{2}$.

First, we begin by evaluating the proposed frequency domain receiver (FDR) performance for a fixed data rate, with different pulse shape. The result is drawn in Fig. 4.

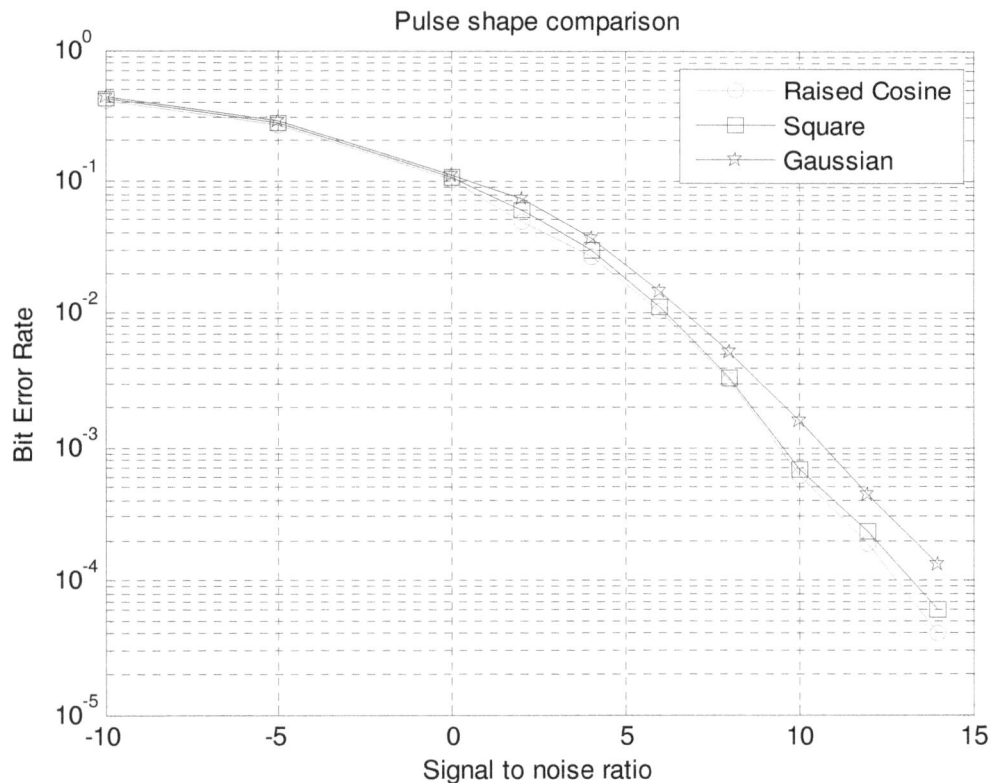

Figure 4: Performance of the FDR receiver for different pulse shape

As shown in Fig. 4, the FDR receiver offers good performance in terms of BER, for the different pulse shapes. We can notice that the square pulse and the raised cosine pulse (with roll off factor α=0.6) offer better performance than the gaussian pulse. In fact, we get an improvment of about $1dB$ when we use the raised cosine pulse. Furthermore, the raised cosine shape is recommended by the FCC because it is the most suitable with the mask imposed by the commission. That's why, we consider the raised cosine pulse as pulse reference in the next simulation.

In wireless communications, there is always a compomise between performance and data rate. That's why, we test the performance of the proposed system for 3 different data rate: 6, 12 and $25Mbit/s$. The results are given in Fig. 5.

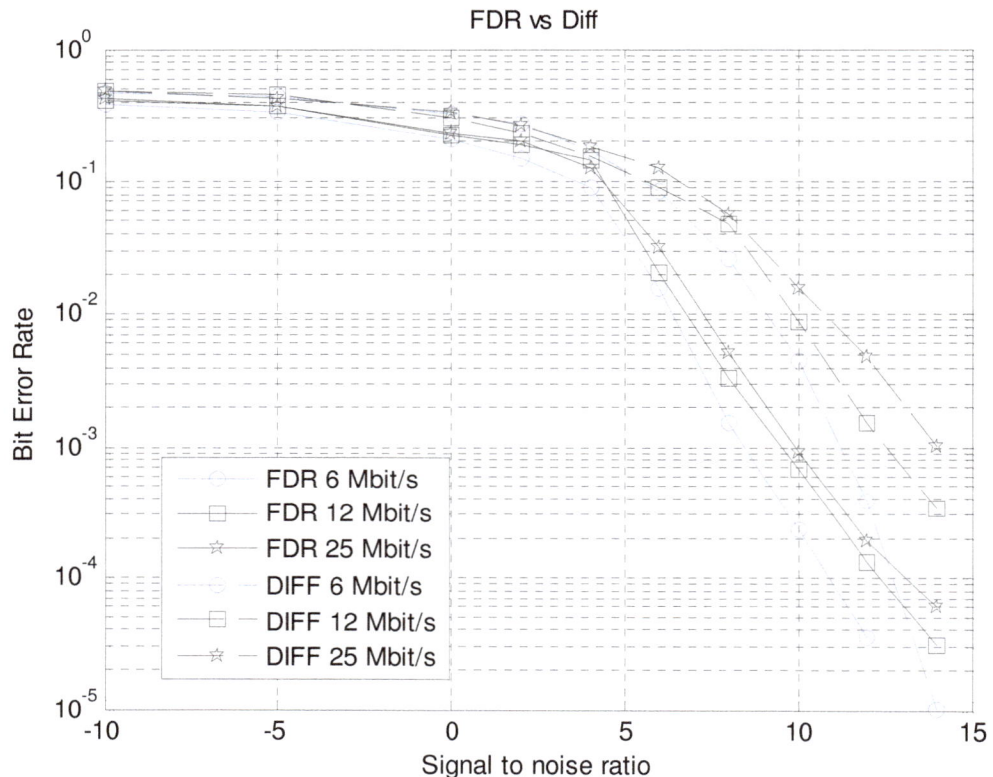

Figure 5: Performance comparision for different data rate

As we can see in Fig. 5 and as expected, the bit error rate increases with the increase of the data rate for both systems: the DS-UWB system using the FDR receiver and the system using the classical differential receiver. But, we can note that both of them are robust to high data rate. On the other hand, the FDR receiver outperforms the differential receiver. For example, for $6Mbit/s$, and to achieve a BER= 10^{-3}, the FDR receiver needs $8dB$ while the differential reciever needs $11dB$, and so we get a gain of about $3dB$. We can also notice for example, a loss of about $2dB$ when increasing the data rate from 6 to $25Mbps$, for the FDR receiver. As for the differential receiver we can perceive a loss of $3dB$.

Thereby, the FDR receiver is more suitable for high data rate.

6 Conclusion

In this paper, a new receiver structure called FDR is proposed for DS-UWB system. It is based on the transformation of the signal arriving at the receiver input from time domain to frequency domain.

We show through simulations that the proposed receiver gives good performance in terms of BER. Besides, the proposed receiver outperforms the differential receiver for the different data rate tested. We have an improvement of about $3dB$ compared to the differential scheme.

As a future work, we propose to evaluate the robustness of the FDR receiver to inter frame interference and multi user interference.

Appendix

The expression of $S_{m,n}(t)$ can be written as:

$$
\begin{aligned}
S_{m,n}(t) &= \sum_{l=0}^{N_f-1} c_l' s_m(l,t) \, exp(j2\pi \frac{nl}{N_f}) \\
&= \sum_{l=0}^{N_f-1} c_l' a_m \, c_i' \, b_m^i \omega_R(t) \, exp(j2\pi \frac{ln}{N_f}) \\
&= \sum_{l=0}^{N_f-1} a_m \, b_m^i \omega_R(t) \, exp(j2\pi \frac{nl}{N_f}), \, (\, c_i' = c_i' \, car \, i = l - mN_f) \\
&= a_m \, \omega_R(t) \sum_{l=0}^{N_f-1} b_m^i \, exp(j2\pi \frac{nl}{N_f})
\end{aligned}
$$

We can write b_m^i as an exponential as follows:

$$b_m^i = exp(j2\pi i \frac{b_m-1}{4})$$

In fact, if $b_m = 1$, we obtain $j2\pi i \frac{b_m-1}{4} = 0$ and so $b_m = exp(0) = 1$

As for the case where $b_m = -1$, we get $j2\pi i \frac{b_m-1}{4} = exp(-j\pi)$ and so $b_m = -1$

Using this tranforamtion of the expression of b_m^i, $S_{m,n}(t)$ becomes:

$$S_{m,n}(t) = a_m. \omega_R(t) \sum_{l=0}^{N_f-1} exp\left[j2\pi \left(\frac{nl}{N_f} + i \frac{b_m-1}{4} \right) \right]$$

From the last expression, we can notice that the value of $S_{m,n}(t)$ depends on the exponential term. That's why, we will try to find the value of this term below. Two cases have to be evaluated: $b_m = 1$ and $b_m = -1$.

- First case: $b_m = 1$

In this case, $S_{m,n}(t)$ is given as follows:

$$S_{m,n}(t) = a_m \, \omega_R(t) \sum_{l=0}^{N_f-1} exp\left[j2\pi \frac{nl}{N_f} \right]$$

- If $n = 0$ then $S_{m,0}(t) = a_m \, N_f \, \omega_R(t)$

- If $n \neq 0$ then $S_{m,n}(t) = a_m\, \omega_R(t)\, \dfrac{1 - exp\left[j2n\pi\frac{lN_f}{N_f}\right]}{1 - exp\left[j2\pi\frac{nl}{N_f}\right]} \Rightarrow S_n(t) = 0$

- Second case: $b_m = -1$

In this case, $S_{m,n}(t)$ can be written as:

$$S_{m,n}(t) = a_m\, \omega_R(t) \sum_{l=0}^{N_f-1} exp\left[j2\pi\left(\frac{nl}{N_f} - \frac{i}{2}\right)\right]$$

- If $j2\pi\left(\dfrac{nl}{N_f} - \dfrac{i}{2}\right) = 0 \Rightarrow \dfrac{nl}{N_f} - \dfrac{i}{2} = 0 \Rightarrow n = \dfrac{l - mN_f}{2l}N_f \Rightarrow n = \dfrac{N_f}{2} - \dfrac{m}{2l}.N_f^2 \in$

$[0, N_f) \Rightarrow n = \dfrac{N_f}{2}$

Hence, if $n = \dfrac{N_f}{2}$, then $S_{m,\frac{N_f}{2}}(t) = a_m\, N_f\, \omega_R(t)$

- If $n \neq \dfrac{N_f}{2} \Rightarrow$

$$S_{m,n}(t) = a_m\, \omega_R(t) \sum_{l=0}^{N_f-1} exp\left[j2\pi\left(\frac{nl}{N_f} - \frac{i}{2}\right)\right]$$

$$= a_m\, \omega_R(t) \dfrac{1 - exp\left[j2\pi N_f\left(\frac{nl}{N_f} - \frac{i}{2}\right)\right]}{1 - exp\left[j2\pi\left(\frac{nl}{N_f} - \frac{i}{2}\right)\right]}$$

We know that:

$$0 \leq n < N_f$$
$$0 \leq \frac{n}{N_f} < 1$$
$$\Rightarrow \qquad 0 \leq \frac{nl}{N_f} < l < N_f$$
$$(0 < i < N_f)$$
$$\Rightarrow \qquad 0 \leq \frac{nl}{N_f} - \frac{i}{2} < N_f - \frac{i}{2}, (i \in [0, N_f))$$
$$\Rightarrow \qquad exp\left[j2\pi N_f\left(\frac{nl}{N_f} - \frac{i}{2}\right)\right] = 1$$

Hence, $S_{m,n}(t) = 0$ if $n \neq \dfrac{N_f}{2}$

If we resume what was said above, $S_{m,n}(t)$ depends on both b_m and n. It is given by expressions below.

- $b_m = 1 \Rightarrow$

$$S_{m,0}(t) = a_m\, N_f\, \omega_R(t)$$
$$S_{m,n}(t) = 0 \ \ if \ \ n \neq 0$$

- $b_m = -1 \Rightarrow$

$$S_{m,\frac{N_f}{2}}(t) = a_m N_f \omega_R(t)$$

$$S_{m,n}(t) = 0 \; if \; n \neq \frac{N_f}{2}$$

Therefore, the decision can be given by $S_{m,n}(t)$ and so $R_{m,n}(t)$. It is a soft decision because $\hat{b}_m$ is determinated by two values of $R_{m,n}(t)$: $n = \frac{N_f}{2}$ and $n = 0$.

First, we suppose that N_f is even.

Since $R_{m,n}(t)$ is complex, we calculate its norm to give the criterion adopted for decision.

$$\left|R_{m,n}(t)\right|^2 = (S_{m,n}(t) + N_{m,n}(t))(S_{m,n}(t) + N_{m,n}(t))^*$$

$$= \left|S_{m,n}(t)\right|^2 + S_{m,n}(t)^* N_{m,n}(t) + S_{m,n}(t) N_{m,n}(t)^* + \left|N_{m,n}(t)\right|^2$$

- If $b_m = 1$ then:

$$\left|R_{m,n}(t)\right|^2 = N_f^2 \omega_R^2(t) + a_m N_f \omega_R(t)(N_{m,n}(t) + N_{m,n}(t)^*) + \left|N_{m,n}(t)\right|^2, if \; n = 0$$

$$\left|R_{m,n}(t)\right|^2 = \left|N_{m,n}(t)\right|^2, if \; n \neq 0$$

- If $b_m = -1$ then:

$$\left|R_{m,n}(t)\right|^2 = N_f^2 \omega_R^2(t) + a_m N_f \omega_R(t)(N_{m,n}(t) + N_{m,n}(t)^*) + \left|N_{m,n}(t)\right|^2, if \; n = \frac{N_f}{2}$$

$$\left|R_{m,n}(t)\right|^2 = \left|N_{m,n}(t)\right|^2, if \; n \neq \frac{N_f}{2}$$

We note that the useful energy is concentrated whether in $R_{m,0}(t)$ if $b_m = 1$, or in $R_{m,\frac{N_f}{2}}(t)$ if $b_m = -1$. As for noise components, they are located in the other cases.

This behaviour is similar to a PPM modulation using frames of duration $N_f T_f$ and pulses having energy N_f times more than the transmitted pulses. And so, the detection is based on an energy collector intergator.

Therefore, to decide of the value of $\hat{b}_m$, we have to use the following criterion:

$$\int_0^{T^{corr}} \left|R_{m,0}(t)\right|^2 dt \geq \int_0^{T^{corr}} \left|R_{m,\frac{N_f}{2}}(t)\right|^2 dt \; \Rightarrow \; \hat{b}_m = 1$$

$$\int_0^{T^{corr}} \left|R_{m,0}(t)\right|^2 dt \leq \int_0^{T^{corr}} \left|R_{m,\frac{N_f}{2}}(t)\right|^2 dt \; \Rightarrow \; \hat{b}_m = -1$$

References

[1] William M. Lovelace and J. Keith Townsend, "The Effects of Timing Jitter and Tracking on the Performance of Impulse Radio", IEEE JOURNAL ON SELECTED AREAS IN COMMUNICATIONS, VOL. 20, NO. 9, DECEMBER 2002.

[2] L.Yang and G. B. Giannakis, "Timing Ultra-Wideband Signals With Dirty Templates", in Proc. IEEE Trans. Commun., VOL.53, no.11, NOVEMBER 2005

[3] Y. Qiao, T. Lv, and L. Zhang, "A new blind synchronization algorithm for UWB-IR systems", in proc. IEEE VTC'09, Spain, April 2009, pp.1–5.

[4] Y. Qiao and T. Lv, "Blind Synchronization and low complexity Demodulation for DS-UWB systems", in proc. IEEE WCNC, 2010.

[5] Z. Tian and G. B. Giannakis, "BER sensitivity to mistiming in ultrawideband impulse radios- part I: Nonrandom channels", IEEE Trans. on Sig. Process., vol. 53, no. 4, pp. 1550–1560, April 2005.

[6] Z. Tian and G. B. Giannakis, "BER sensitivity to mistiming in ultrawideband impulse radios- part II: fading channels", IEEE Trans. on Sig. Process., vol. 53, no. 5, pp. 1897–1907, May 2005.

[7] J. R. Foerster and Q. Li, "Uwb channel modeling contribution from intel",IEEE 802.15.3 Wireles Personal Area Networks, Tech. Rep. IEEE p802.15-02/279r0-SG3a,Jun. 2002.

[8] N. Boubaker and K. B. Letaief, "Ultra Wideband DSSS for Multiple Access Communications Using Antipodal Signaling", Communications, 2003. ICC '03. IEEE International Conference on, Volume: 3, On page(s): 2197- 2201 vol.3

[9] L. Zheng, M. Mdard, D.N.C. Tse and C. Luo, "On Channel Coherence in the Low SNR Regime, 41st Allerton Annual Conference on Communication", Control and Computing, October 2003

[10] Gezici S., Tufvesson F. and Molisch A.F., "On the performance of transmitted-reference impulse radio", Global Telecommunications Conference, 2004. GLOBECOM '04. IEEE Volume 5, 29 Nov.-3 Dec. 2004 Page(s):2874 - 2879 Vol.5

[11] Pausini M. and Janssen G.J.M.; "Analysis and comparison of autocorrelation receivers for IR-UWB signals based on differential detection", Acoustics, Speech, and Signal Processing, 2004. Proceedings. (ICASSP '04). IEEE International Conference on, Volume 4, 17-21 May 2004 Page(s):iv-513 - iv-516 vol.4

[12] Witrisal K. and Pausini M.; "Equivalent system model of ISI in a framedifferential IR-UWB receiver", Global Telecommunications Conference, 2004. GLOBECOM '04. IEEE, Volume 6, 29 Nov.-3 Dec. 2004 Page(s):3505 - 3510 Vol.6

[13] Durisi G. and Benedetto S., "Performance of coherent and noncoherent receivers for UWB communications", Communications, 2004 IEEE International Conference on Volume 6, 20-24 June 2004 Page(s):3429 - 3433 Vol.6

[14] A. A. Boudhir, M. Bouhorma, M. Ben Ahmed and Elbrak Said,"The UWB Solution for Multimedia Traffic in Wireless Sensor Networks", International Journal of Wireless & Mobile Networks, October 2011.

[15] Susmita Das and Bikramaditya Das,"A comparision study of time domain equalization technique using uwb receivers performance for high data rate WPAN system", International Journal of Computer Networks \& Communications (IJCNC), Vol.2, No.4, July 2010.

[16] Adel A. M. Saleh et REINALDO A. Valenzuela, "A Statistical Model for Indoor Multipath Propagation", IEEE Journal on selected Areas in COMMUNICATIONS. VOL. SAC-5. NO. 2. FEBRUARY 1987.

RACH CONGESTION IN VEHICULAR NETWORKING

Ramprasad Subramanian[1] and Kumbesan Sandrasegaran[2]

[1&2]Centre for Real-time Information Networks,
School of Computing and Communications, Faculty of Engineering and Information
Technology, University of Technology Sydney, Sydney, Australia

ABSTRACT

Long term evolution (LTE) is replacing the 3G services slowly but steadily and become a preferred choice for data for human to human (H2H) services and now it is becoming preferred choice for voice also. In some developed countries the traditional 2G services gradually decommissioned from the service and getting replaced with LTE for all H2H services. LTE provided high downlink and uplink bandwidth capacity and is one of the technology like mobile ad hoc network (MANET) and vehicular ad hoc network (VANET) being used as the backbone communication infrastructure for vehicle networking applications. When Compared to VANET and MANET, LTE provides wide area of coverage and excellent infrastructure facilities for vehicle networking. This helps in transmitting the vehicle information to the operator and downloading certain information into the vehicle nodes (VNs) from the operators server. As per the ETSI publications the number of machine to machine communication (MTC) devices are expected to touch 50 billion by 2020 and this will surpass H2H communication. With growing congestion in the LTE network, accessing the network for any request from VN especially during peak hour is a big challenge because of the congestion in random access channel (RACH). In this paper we will analyse this RACH congestion problem with the data from the live network. Lot of algorithms are proposed for resolving the RACH congestion on the basis of simulation results so we would like to present some practical data from the live network to this issue to understand the extent RACH congestion issue in the real time scenario.

KEYWORDS

RACH; Congestion; LTE; Human to Human (H2H);Machine to Machine (M2M);Vehicle Nodes (VN); Mobile ad hoc network (MANET);Vehicular ad hoc network (VANET)

1. INTRODUCTION

Enabling wireless connectivity to the cars and making the transportation system intelligent has become a buzz word in lot of developed/developing countries. The Global car sales by 2020 is projected to be around 90 million units. The trend is going to go upwards in the subsequent years. As per the survey conducted by Bureau of Transport and Regional Economics in Australia[1], the projected travelling distance by people would be around 275 billion kilometres. So this raises another important point that people are going to spend lot of time in cars than at home and other place. So it become imperative that we must extend all the facilities of modern communication systems to the cars. So in this aspect vehicular communication is becoming more essential day by day. On the other hand the growth of communication systems has seen some tremendous growth in terms of technology and as well as with customer base. Moreand more people are migrating from GSM/3G to LTE because of high data rate. With the advent of LTE/LTE-A the data usage has surpassed the other traditional services like voice and text. In a recent study by Deloitte by 2016 LTE is forecast to carry more data traffic than 3G globally [2]. The imperative for carriers will be to build coverage and capacity as quickly and economically as possible. So when we carefully analyze this situation the extension of this communication facility inside the car environment and making our transportation system intelligent would be the ideal step forward.

In order to implement intelligent transportation system (ITS) IEEE has specified some standards like IEEE 802.11 p[1] which supports VANETs. This technology is very easy to deploy, cost effective and mature technology. But comes with some disadvantage like scalability issue, Quality of service (QoS) guarantees and it does not have proper end to end infrastructure for wide coverage. On the other hand LTE provides adequate infrastructure support and hence can guarantee QoS, it can provide wide coverage and the issue of blind spots (no coverage area zones) is minimized. But LTE comes with another disadvantage like network congestion because of growing customer base. So in this paper we will analyze the problem of congestion in LTE network which acts as an impediment to cater the ITS applications and service to the vehicles.

2. FUNDAMENTAL CONSIDERATION

The basic idea of 2G, 3G and 4G architecture design is to serve H2H communication. But with growing competition between the various mobile operators to capture the major chunk of the customer base the operators are forced to increase the investment in network expansion, QoS and low cost services and hence results in reduced average revenue per user (ARPU). This naturally brings down the capex and increases opex. So in order to increase the revenues the operators are looking for various avenues to make profits. One such avenue is providing latest communication services inside the VNs. The VNs can also be categorized under M2M or MTC devices. This service can increase the mobile operators' connection and revenue growth and as well as the most common go-to-market scenarios that apply to mobile operators in the M2M value add chain.

2.1. LTE network architecture

LTE, unlike its predecessor technologies like 2G and 3G, LTE is designed completely to provide seamless internet protocol connectivity between the user equipment (UE) and packet data network (PDN). While the term "LTE" includes the evolution of the universal mobile telecommunications system (UMTS) radio access through the evolved UTRAN (E-UTRAN), it is also accompanied by an evolution of the term "System Architecture Evolution" (SAE), which encompasses the evolved packet core (EPC) network. Together the LTE and SAE comprise the evolved packet system (EPS). EPS uses the EPS bearers to route the IP traffic from the gateway in the PDN to the UE. A bearer can be defined as an IP packet flow with a defined QoS between the gateway and the UE. The set up and release of bearers with respect to the applications are provided by the E-UTRAN and EPC together. EPS provides the user with IP connectivity to a PDN for internet accessing, and for running services such as Voice over IP (VoIP). An EPS bearer is typically associated with a QoS. Multiple bearers can be established for the user in order to provide multiple QoS streams or connectivity to different PDNs. For example, a user can be engaged in a voice (VoIP) call and can perform web browsing or FTP download at the same time. The necessary QoS for the voice call would be provided by the VoIP bearer, while for web browsing and FTP session the necessary QoS would be provided by the best-effort bearer. The LTE network architecture is designed in such a way to provide sufficient security protection for the network and privacy to the users against the fraudulent usage of the network.

At a high level, the LTE network comprises of the Core Network (CN) which is EPC and the access network E-UTRAN. While the CN is made up of many logical nodes, the access network is comprises of just one node, the evolved NodeB (eNB), which in turn connects to the UEs. Every network elements in the LTE architecture is interconnected by various interfaces which are standardised by the 3GPP in order to allow interoperability between different vendors. This gives the opportunity to the network operators to source different network elements from different vendors. Based on this the network operators have the freedom to choose between the vendors and they can construct their network with a single vendor or they can split choose the vendors for various network elements. depending on commercial considerations.

The CN (called as evolved packet core in SAE) is responsible for the overall control of the network and the establishment of the bearers various services. The main logical nodes of the EPC are:

• PDN Gateway (P-GW)
• Serving Gateway (S-GW)
• Mobility Management Entity (MME)

In addition to those above specified nodes, EPC also includes other logical nodes and functions such as the Home Subscriber Server (HSS) and the Policy Control and charging Rules Function (PCRF). Since the EPS only provides a bearer path of a certain QoS services, the control of multimedia applications such as VoIP is provided by the IP Multimedia Subsystem (IMS), which is considered to be outside the EPS itself.

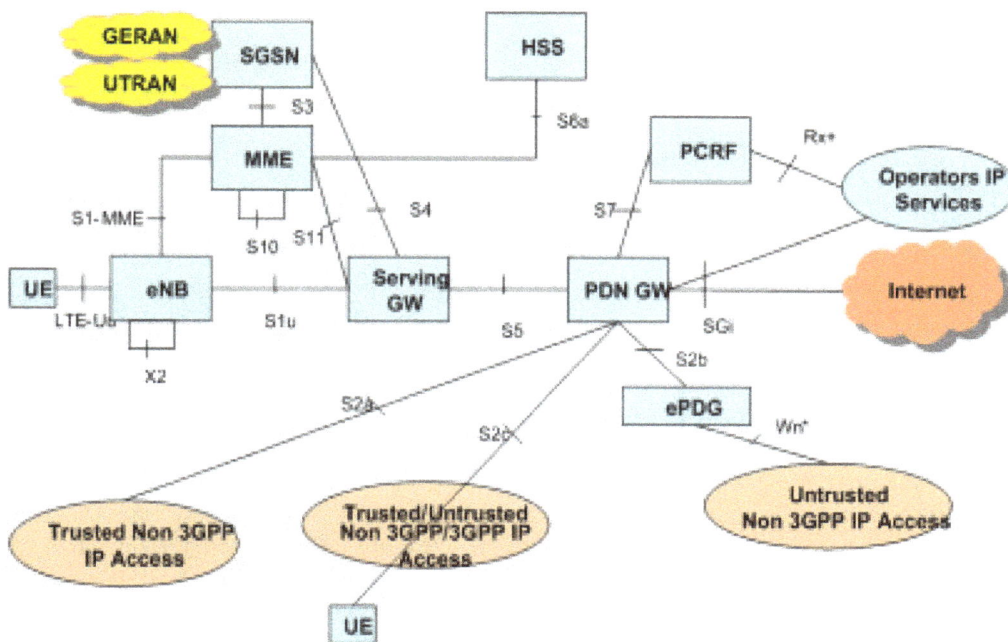

Figure 1. LTE Architecture

Table 1. LTE reference points.

PCRF	Responsible for policy control decision-making, as well as for controlling the flow-based charging functionalities in the Policy Control Enforcement Function (PCEF), which resides in the P-GW.
HSS	The Home Subscriber Server contains users' SAE subscription data such as the EPS-subscribed QoS profile and any access restrictions for roaming.
P-GW	The PDN Gateway is responsible for IP address allocation for the UE, as well as QoS enforcement and flow-based charging according to rules from the PCRF.
S-GW	All user IP packets are transferred through the Serving Gateway, which serves as the local mobility anchor for the data bearers when the UE moves between eNodeBs.
MME	The Mobility Management Entity (MME) is the control node that processes the signaling between the UE and the CN.

2.2. LTE infrastructure as Backbone in Vehicular Networking

There are several reasons why LTE is considered as one of the leading contender to provide backbone infrastructure for vehicular networking.

The first and foremost reason for this is adequate coverage of LTE network. As the subscriber base increases every day the operators are rapidly expanding the LTE network. This in turn provides good coverage for vehicle networking applications. The second reason, is the LTE provides good QoS in terms of data throughput in the downlink channel. So if a user of the vehicle nodes makes a request which amounts to some big bandwidth requirement then LTE can readily support it. Thirdly, LTE network provides a centralised architecture which can be used by the vehicle networking applications to reach the central content server to request or make fresh demand. The LTE air interface can support various access technologies like time-division duplexing (TDD), frequency- division duplexing (FDD), and half-duplex FDD schemes; it also provides channel bandwidth of (1.4–20 MHz). Concerning the access technology in LTE, orthogonal frequency-division multiple access (OFDMA) is used in the downlink and single-carrier frequency-division multiple access (SC-FDMA) in the uplink, both providing high flexibility in the frequency-domain scheduling. Usage of multiple-input multiple-output (MIMO) techniques in LTE would improve the spectral efficiency by a factor of 3 to 4 compared to other generation (2G/3G) systems even at very high speeds, making LTE very efficient in challenging and dynamic propagation environments like the vehicular one. eNodeB manages the radio resources centrally at every transmission time interval of 1ms duration and provides efficient QoS while increasing the channel utilization. Packet scheduler plays an important role in the eNodeB by selecting the traffic flow to serve the UEs based on the QoS requirements and services (as specified by the QoS class identifier, QCI), and further decides the suitable modulation and coding scheme based on feedback from the mobile terminals about the channel in the channel quality indicator (CQI). QCI refers to a set of packet forwarding treatments, for example, resource type (guaranteed or not guaranteed bit rate), priority, packet loss rate and delay budget. LTE also supports high-quality multicast broadcasting and multicasting services which is similar to 3G but it is evolved in LTE and termed as eMBMS in the core and in the radio access network. It offers the possibility of sending the data *only once to a set of users* registered to the offered service, instead of sending it to *every node separately*. The standardization of LTE-Advanced (LTEA) is ongoing in 3GPP (Rel. 11) as a major enhancement of LTE in terms of bit rate, capacity, and spectral efficiency, mainly through the support of advanced MIMO techniques, carrier aggregation, and relay nodes. With LTE-A still in an early stage, the focus of the related work reported in this article is on LTE.

Since LTE provides centralized architecture it cannot support Vehicle to vehicle (V2V) communications[3][4]. But there is also some disadvantage of LTE being used in this vehicular application. The first and foremost is, in cellular network the non active devices will be in idle mode in-order to save the network resources. So in-order to transmit the request to the central content server the VNs should be in active mode. The VNs should request for random access channel (RACH) to request for resources to transmit some data in uplink channel (PUSCH). But because of RACH overloading/congestion in the LTE network the VNs may experience collision in preamble ID and can face problems to receive random access response (RAR) from the eNB. The focus of this paper is to prove that in future with growing number of LTE subscribers and also equally increasing MTC devices the Rach congestion/overloading [5] is going to be a big bottleneck for these devices to attach to the network. Many people have published various algorithms for this RACH congestion issue on the basis of computer simulation results. But in this paper an attempt is made to analyse the RACH data from the real network which was collected for over the period of three months. This was supported by set simulations which was done to analyse the preamble ID collision and re-transmission issue.

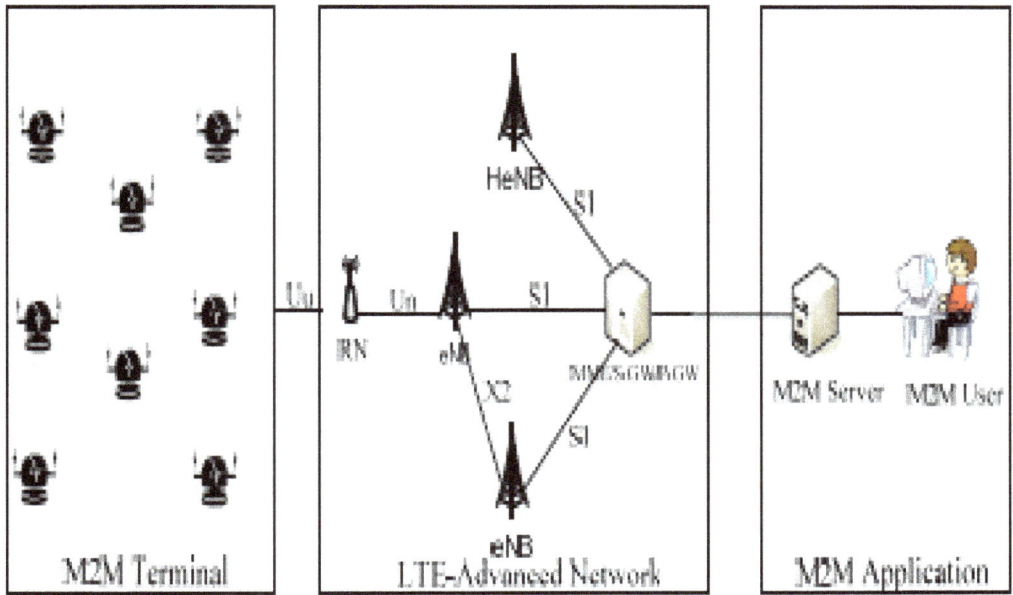

Figure 2. Vehicular networking architecutre (Anthony Lo et al.,)

3. Problem Statement

With ever increasing vehicle population and subsequent increase in road infrastructure to support it, the transportation scenarios has become complex day by day. The government agencies and infrastructure planners are working more seriously to plan for future traffic growth. For example a small incident in a road can cause huge traffic blocks and delays. So in this scenario the information regarding the roads needs to be published to the drivers so that they can be diverted to their destinations through other available alternative routes. Lot of techniques are employed to enable this to happen. One such technique is by using by LTE in vehicular networking. As per the ETSI publications[6] the number of MTC devices are expected to touch 50 billion by 2020 and this will surpass the H2H. With growing congestion in the LTE network, accessing the network for any request from vehicles (from idle mode to active mode) especially during peak hour is going to be a big challenge because of RACH congestion/overloading. In a LTE cell 64 preambles signatures are available in that some are used for contention based allocation and some are used for contention free allocation. So not all 64 preamble signatures are available for UE's or VNs to use to access the network this results in collision and re-transmission which clogs the network. This will have a serious impact in the freshness of the data received from the network.

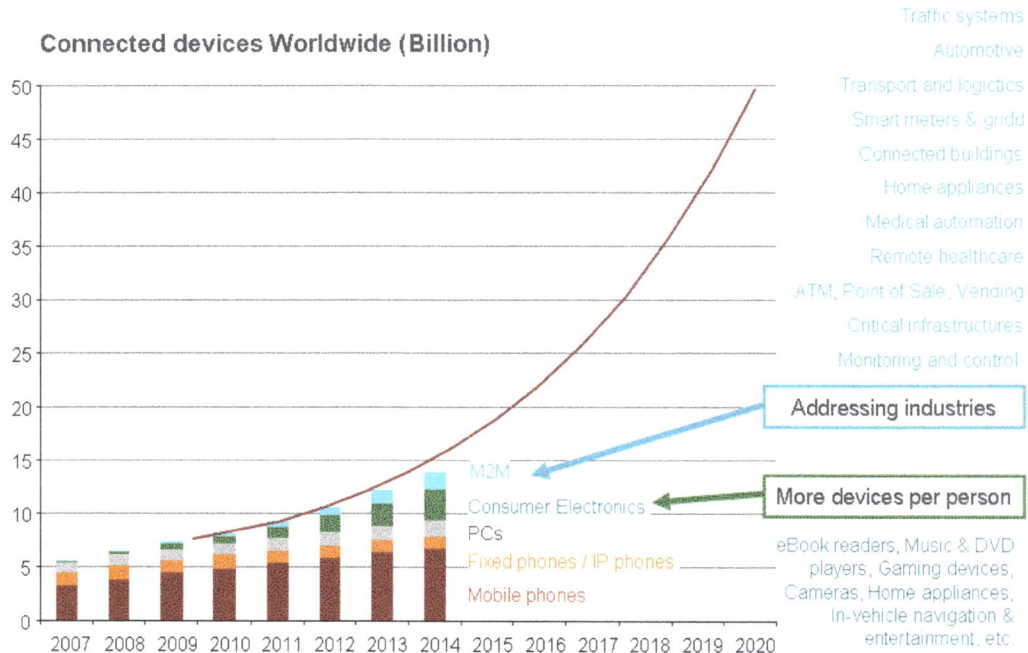

Figure 3. M2M devices service and forecast (Ericsson)

3.1. RACH process and Congestion - Simulation Results

The RACH procedure begins when VNs has something to transmit on the physical random access channel (PRACH). The process starts the message exchange between the VNs and eNB. This process has two different types of variants and they are contention based random access procedure and contention free random access procedure. The later part is used by the network to assign a preamble to the UE during the handover process and since the preamble IDs are handled by the network the congestion in this procedure is virtually non-existence. But the focus of this paper is contention based random access procedure. The VNs are usually in idle mode and they will not be in connect mode to the network. So when a request is being placed in the VNs they contact the eNB for resource allocation in uplink. As a result of this the VNs will receive resource to transmit the query in the uplink, initial timing advance value in the uplink and temporary C-RNTI.

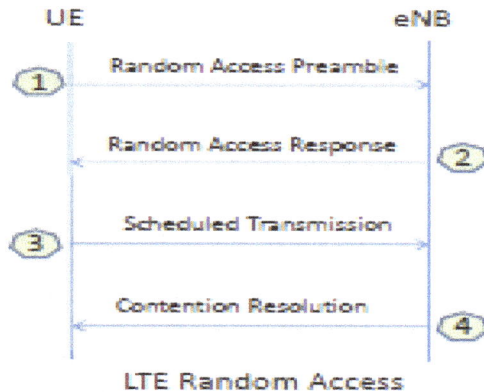

Figure 4. Random access procedure in LTE (Han-Chuan Hsieh et. al.,)

Then status of the VN will change to RRC_IDLE to RRC_CONNECTED when the VN wishes to access the network to submit a query. But it will not have a PUSCH resource in the uplink channel to send the request. So it triggers a random access procedure. The VNs reads the cells random access configuration from SIB2 and chooses a preamble sequence at random from the available ones (apart from the ones used by the system for the contention based scheme). It then transmit on the same resource blocks using the same preamble sequence. The base station sends the scheduling command in RAR and this will be followed with the schedule transmission by the VN to the eNB in this process the VNs identifies itself using either by S-TMSI or by random number. The eNB sends a acknowledgement using PHICH. The base station address the VN using C-RNTI which it allocated earlier. It follows the command with MAC control element called contention resolution identity.

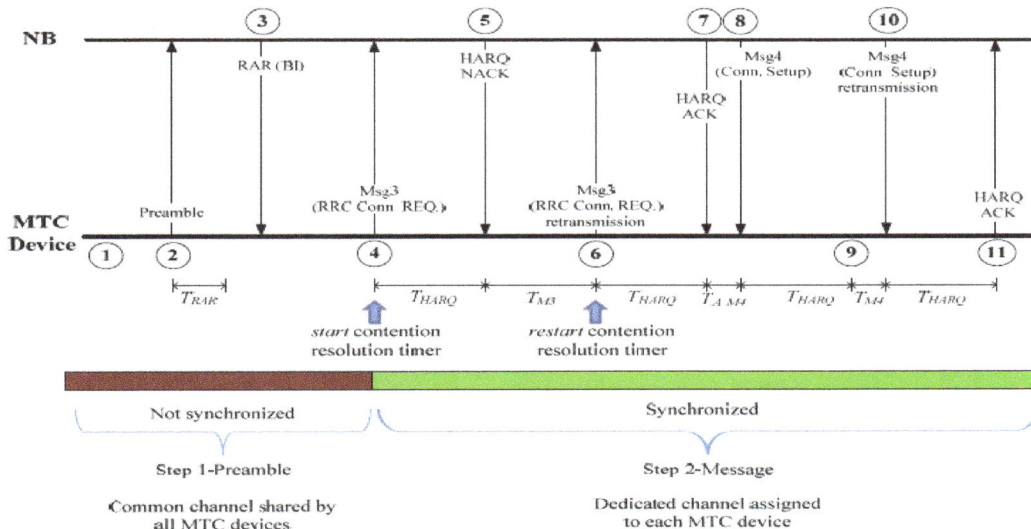

Figure 5. Random access procedure steps in VNs/MTC devices (Chia-hung wei et al.,)

3.2. RACH Overload/Congestion - Simulation Results

The overload control of uplink random-access channel (RACH) in radio access network (RAN) is one of the principle working items for 3GPP Long Term Evolution. The purpose of RAN overload control is to avoid RAN overload when mass VNs devices simultaneously contend for the RACH. From the perspective of the way that VNs/MTC traffic is generated, the RAN overload control schemes can be categorized into push-based and pull-based approaches. In the push-based approach, the VNs/MTC traffic is *pushed* from VNs/MTC devices to the network without any restriction until RAN overload is detected this scheme is not controlled by the network. In the pull-based approach, the VNs/MTC traffic is *pulled* by the network and thus, the network may properly control the VNs/MTC traffic load through paging and thus, prevents RAN overload. In another scheme called access class barring (ACB), separate RACH resources are allocated for VNs/MTC, in this scheme as dynamic allocation of RACH resource takes place, in VNs/MTC backoff scheme the VNs will be told to backoff if there is heavy congestion in the network and finally in slotted access scheme each VNs will be allocated a certain time slots and the VNs are allowed to transmit only during that time. In LTE, a downlink paging channel is defined to transmit the paging information to user equipment (UE), informing UEs on system information changes and emergency notifications. The network may transmit a paging message to activate a specific UE at the UE's paging occasion. The paging occasion of each UE is determined according to its UE identity (UE-ID). Current paging mechanism that was originally designed for H2H services can only page up to 16 devices with a single paging message, and only two paging occasions are available per 10 ms radio frame. Therefore, a BS must transmit multiple paging messages over a long period to activate a large number of VNs/MTC devices which substantially increases the load in the BS.

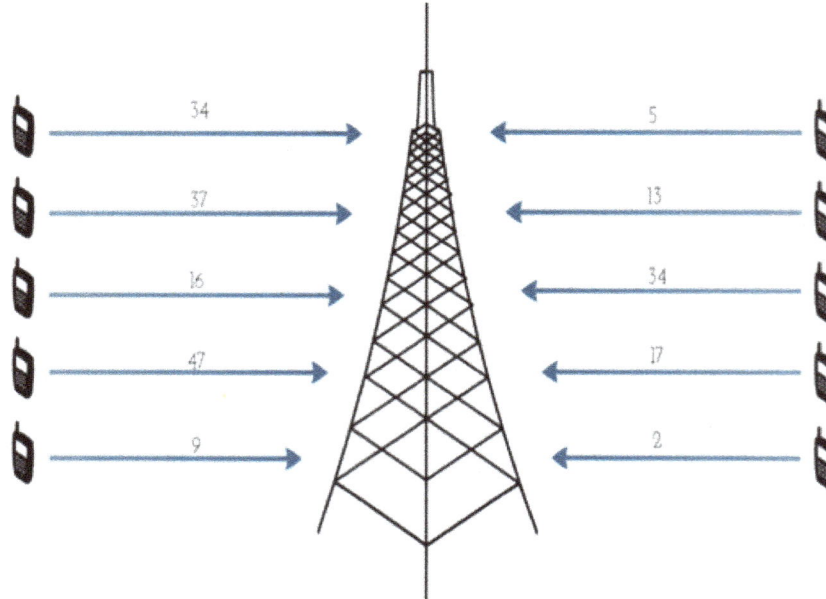

Figure 6. Random access procedure attempt with Preamble ID by UE/VNs

If VNs transmits same random number in different resource blocks there will not be any contention but if VNs transmit the same random number in the same resource blocks then there will be a contention. So the devices go for re-transmission automatically until it reaches the max retransmission attempts This scenarios likely to happen almost in all the network. The VNs and mobiles are going to use the same resources to access the network. As such the networks are busy because of the increasing customer base in LTE and as projected by ETSI many more MTC devices are going to join the same bandwagon in the future.

As per the simulation results developed using NS3[7] the probability of preamble collision increases as the number of mobile devices increases in the LTE cell.

Table 2. Nodes and collision percentage.

No of nodes attempting for Rach	Collision percentage
10	20
40	37.5
100	58
200	75

Table 3. Number of nodes and percentage of successful preamble id throughput.

No of nodes attempting for Rach	Percentage of Successful Preamble ID throughput
10	80
40	62.5
100	42
200	25

Table 4. Collision percentage and re-transmission attempts

Collision percentage	Retransmission attempts
20	3
37.5	10
58	25
75	50

The preamble collision and re-transmission simulation results shown in table 3 as the number of nodes increases the percentage of collision also increases and the re-transmission to make it through (table 4). This does not mean that the attempt is an failure. But the main disadvantage of increase in the number of re-transmission will increase the delay for the VNs to get the resource in the uplink channel to access the network for any request which will have adverse effect in the freshness of the data received and also will result in network wide congestion in the RACH.

3.2. Expected number of Unsuccessful users in accessing the RACH

The above simulation results confirms the percentage of collision and the number of re-transmission attempts to access the network. But that doesn't confirm that the attempts are failure ones. So in order to analyse the number of failures in accessing the uplink resource to access the network a practical approach was taken instead of simulation routes. The data that has been presented in figure 6 was the result of collection of counters for RACH failures for past three months from an live LTE network and in which more than a billion of RACH attempts where studied and from that RACH failure rate has been calculated. The below results shows only the RACH failures in the network. From this below figure we can attribute that 30% of the RACH attempts in the networks is failing and only 70% of the RACH attempts are successful. Through this analysis we want to confirm that LTE network is already getting clogged up without much use of VNs or MTC devices as of today and the situation will get worse if we start using them. At the same time a proper random access technique should be addressed to improve the situation as the current techniques has many shortcomings.

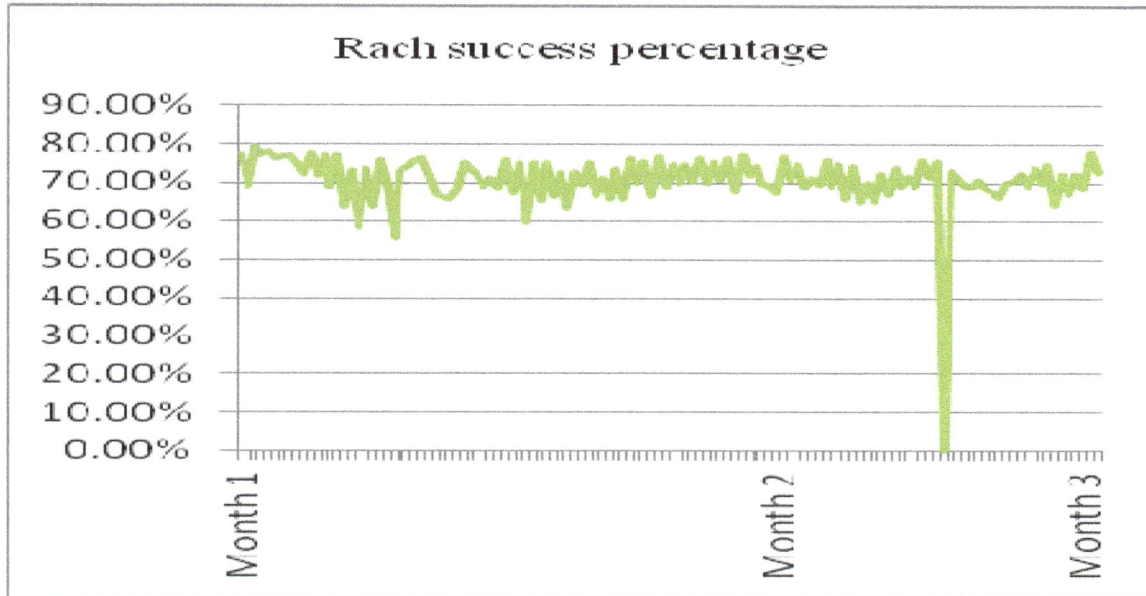

Figure 6. Rach failures results from Live network

4. CONCLUSION AND FUTURE WORK

As the multiscale framework to ITS modelling, network architecture design and traffic control of ITS is being developed with all the facilities of modern communication should be extended to vehicle environment. But in order to do so various technical challenges needs to addressed. One of the major challenge is to analyze the impact on the preamble collision, retransmission and RACH failures of the LTE network. In that attempt firstly, simulation was used to scale the preamble collition and re-transmissions was made and secondly, the data from the live network which was collected for over three months were used to scale the RACH failures. The results obtained will be used to design the new random access technique for M2M communication. So that the M2M services can co-reside along with the H2H services.

REFERENCES

[1]   Bureau of Infrastructure, Transport and Regional Economics -BITRE, (2009) "Greenhouse gas emissions from Australian transport: projections to 2020", Working paper 73, 2009, Canberra ACT.

[2]   Deloitte survey, (2013)   "State of the global mobile consumer - 2013 divergence deepens", www.deloitte.com/globalmobile2013.

[3]   Giuseppe Araniti, Claudia Campolo, Massimo Condoluci, Antonio Iera, and Antonella Molinaro, (2013) "LTE for Vehicular Networking: A Survey", IEEE Communications Magazine, May 2013, pp 148 - pp157.

[4]   Min Chen, Jiafu Wan and Fang Li, (2012) "Machine-to-Machine Communications: Architectures, Standards and Applications", Transactions on Internet and Information Systems, vol. 6, no. 2, February 2012.

[5]   3GPP TS 22.368 V11.3.0, (2011) "Service requirements for Machine-Type Communications (MTC)" Stage 1, September 2011.

[6]   ETSI, (2011) "Standards on Machine to Machine Communications", Mobile world congress, Barcelona.

[7]   NS-3: Simulator, http://www.nsnam.org/

Authors

Ramprasad Subramanian is an experienced telecom engineer in the field of 2G/3G and LTE/LTE-A. He holds M.S (By research) in Information and Communication from Institute of Remote Sensing, Anna University (India)(2007) and Bachelors of Engineering in Electronics and Communication engineering from Bharathidasan University (2001)(India). He has done many projects in the area of 2G/3G and LTE. He has done many consultative projects across Africs/Americas/Asia etc. He was the recipient of India's best invention award for the year 2004 from Indian Institute of Management Ahmadabad and Government of India. His current research focuses on 4G mobile networks and vehicular Ad hoc networks.

Kumbesan Sandrasegaran is an Associate Professor at UTS and Director of the Centre for Real-Time Information Networks (CRIN). He holds a PhD in Electrical Engineering from McGill University (Canada)(1994), a Master of Science Degree in Telecommunication Engineering from Essex University (1988) and a Bachelor of Science (Honours) Degree in Electrical Engineering (First Class) (1985). He was a recipient of the Canadian Commonwealth Fellowship (1990-1994) and British Council Scholarship (1987-1988). His current research work focuses on two main areas (a) radio resource management in mobile networks, (b) engineering of remote monitoring systems for novel applications with industry through the use of embedded systems, sensors and communications systems. He has published over 100 refereed publications and 20 consultancy reports spanning telecommunication and computing systems.

13

A LOW-ENERGY FAST CYBER FOR AGING MECHANISM FOR MOBILE DEVICES

Somayeh Kafaie[1], Omid Kashefi[1] and Mohsen Sharifi[1]

[1]School of Computer Engineering,
Iran University of Science and Technology, Tehran, Iran
`so_kafaie@comp.iust.ac.ir, kashefi@{ieee.org, iust.ac.ir},`
`msharifi@iust.ac.ir`

ABSTRACT

The ever increasing demands for using resource-constrained mobile devices for running more resource intensive applications nowadays has initiated the development of cyber foraging solutions that offloadparts or whole computational intensive tasks to more powerful surrogate stationary computers and run them on behalf of mobile devices as required. The choice of proper mix of mobile devices and surrogates has remained an unresolved challenge though.In this paper, we propose a new decision-making mechanism for cyber foraging systems to select the best locations to run an application, based on context metrics such as the specifications ofsurrogates, the specifications of mobile devices, application specification, and communication network specification. Experimental results show faster response time and lower energy consumption of benched applications compared to when applications run wholly on mobile devices andwhen applications are offloaded to surrogates blindly for execution.

KEYWORDS

Pervasive Computing, Cyber Foraging, Offloading, Surrogate,Mobile Computing, Resource Constraint.

1. INTRODUCTION

Nowadays mobile devices are very popular and people all over the world are increasingly using mobile devices such as cell phones and PDAs to run many applications from daily tasks to emergencies. Finally, using mobile devices and wireless networks, accessing information anywhere and anytime seems more achievable[1].In recent years, users benefit from mobile devices to use more resource intensive applications. Some examples of such applications are natural language translator [2, 3], speech recognizer [2, 3], optical character recognizer [2], image processor [4], and games with high amount of processing [5].

However, there are often shortcomings in quality of mobile devices' tasks due to their resource poverty. The mentioned applications require higher computing power, memory, and battery lifetime than is available on resource constrained mobile devices. They also require faster responses than is currently supported on mobile devices. Unfortunately, at any level of cost and technology, considerations such as weight, size, battery life, ergonomics, and heat dissipation impose severe restrictions on computational resources such as processor speed, memory size and disk capacityof these devices[6]. Therefore, mobile devices always remain more resource constrained than traditional stationary computers [6, 7].

On the other hand, a pervasive computing environment is an environment that focuses on mobility and usage of mobile devices [8]. Pervasive computing was first introduced by Mark Weiser [9] in 1991;Satyanarayanan[10] has defined pervasive environments as "environments saturated with computing and communication capability, yet gracefully integrated with human users".

One of the most important and favourable solutions to cope with resource poverty of mobile devices, especially in pervasive computing, is *cyber foraging*. Generally, cyber foraging is task offloading in order to resource augmentation of a wireless mobile device by exploiting available static computers [10]. In cyber foraging approach, the mobile device sends the whole or a part of an application to nearby idle static computers, called *surrogate* and receives the results to improve the response time and/or accuracy, or confront with its resource constraint. In this paper, we study effectiveness of cyber foraging from mobile devices, surrogates, application, and network aspects.

The remainder of the paper is organized as follows. Related researches on task offloading are discussed in Section 2. In Section 3, the mobility constraints, cyber foraging idea and effectiveness of this idea to alleviate the constraints are explained. Section 4 presents our proposed cyber foraging approach. The results of experimental evaluations are depicted and discussed in Section 5, and Section 6 concludes the paper.

2. RELATED WORK

There are several approaches with different objectives that have used the offloading of applications, but the term "cyber foraging" was first introduced by Satyanarayanan[10]. Cyber foraging is the discovery of static idle computers called surrogates in the vicinity of a mobile device and entrusting some of the tasks of the mobile device to them [10]. As computers become cheaper and more plentiful, cyber foraging approaches become more reasonable to employ.

Spectra [3]is the first cyber foraging system that isfocused on reducing the latency and energy consumption.Spectra adds a feature called *self-tuning* to monitor application behaviour and estimate the resource demandof an application. Spectra's approach to measure energy consumption of the tasks does not work well enough, in some cases. Furthermore developers must follow most of the cyber foraging steps in Spectra manually that it causes significant changesin the code.

Chroma [2, 11, 12]is an extension of Spectra which tries to improve it by reducing the burden on developers. To do so, Chroma uses a new concept called *tactics* that are meaningful ways of application partitioning, specified by the programmer.Chroma uses a fixed utility function to improve latency but ignores battery lifetime.Furthermore, Chroma presents three ways applicable in environments that are full of idle computing resources.First it sends a task execution request to several surrogates in paralleland chooses the fastest response; second it splits operation data and forwards each part to a different surrogate;third it sends the same task execution request with different quality to different surrogates and picks the result with the highest quality that satisfies the latency threshold.

On the other hand,Gu*et al.*[13] have used a graph model to select offloading parts of the program to improve memory constraint of mobile device.Ou*et al.*[14, 15]have expanded their approach and have used a similar method to address the CPU and bandwidth constraint, too. Song *et al.*[16, 17] has proposed a middleware architecture, called MobiGo, for seamless mobility to choose the best available service according to the bandwidth and latency, and Kristensen[18, 19] has introduced Scavenger as a cyber foraging framework whose focus is on CPU power of mobile devices.

However, none of the mentioned works, except Spectra, address directly energy constraint in mobile devices. Othrnan*et al.*[20] were one of the oldest researchers who employ the offloading to reduce the power consumption. Kemp *et al.*[21] also presented Ibis to compare offloading with local execution in terms of responsiveness, accuracy and energy consumption.Cuervo*et al.*[22] present an infrastructure, called MAUI to offload the applications and reduce the energy consumption. MAUI supports programs written in managed code environments such as Microsoft .Net CLR and Java. It provides a graph of program's methods and divides them into local and remote groups to execute.They have located the solver (decision-making unit) out of the mobile device to decrease the computation cost, while burden more communication cost.

In this paper, we propose a context-aware decision-making mechanism to make decisions about task offloading in terms of not only energy consumption, but also current processing power and available memory to improve response time and energy consumption in mobile devices.

3. CYBER FORAGING AND MOBILE COMPUTING

3.1. Augmented Mobile Devices

Mobile devices, due to their mobility nature, cannot be plugged in most of times. Therefore, energy consumption is one of the most important constraints of mobile devices [23]. On the other hand, portability requirements necessitate being as light and small as possible. The inherent constraints include low processor speed, memory, storage size, network bandwidth and limited battery lifetime.

Ubiquitous availability of advanced mobile technologies makes users to expect to run the same applications on mobile devices and static computers. However, regarding resource poverty of mobile devices, it is evident that static computers perform the tasks faster and more accurate. Besides, it is possible that the mobile device does not have sufficient memory, storage or battery to complete the task.

To run the task on a static computer (i.e. surrogate) on behalf of the mobile device, it is required to send the related code and data from the mobile device to the surrogate and receive back the results, which is a time and energy-consuming process. The time of sending/receiving data (application code, input parameters and results) to/from the surrogate depends on the size of data and results as well as on the network bandwidth.

Cyber foraging causes reduction of execution time and energy consumption due to the exploiting more powerful surrogates, but transmission of associated information increases response time and decreases battery lifetime. Since communication usually consumes more energy than computation [21], it raises an issue: "under which circumstances is it worth to use offloading?". Therefore, a decision system must imply that a task is worth to offload to a surrogate or not. In this paper, we present a mechanism to decide about task offloading according to the context information.

3.2. Cyber Foraging Steps

A cyber foraging approach includes some steps that every available cyber foraging systems have considered all or some of them. These steps can be summarized as follows.

- *Surrogate discovery.* First of all, available idle surrogates that are ready to share their resources with the mobile device must be found. Some researches [13, 18] have addressed surrogate discovery.

- *Context gathering*. To have a good decision about target execution location, there is a need to monitor available resources in surrogates and mobile devices and estimate application resource consumptions which isconsidered as context gathering in some cyber foraging systems [2, 3, 18].

- *Partitioning*. In this step, a task is divided into smaller size subtasks, and undividable i.e. unmovable parts are specified. Some researches [24] do the partitioning automatically.

- *Scheduling.*The most important step of cyber foraging is to place each task at the surrogate(s) or the mobile device most capable of performing it, based on the context information and the estimated cost of doing so. Many researches[3, 12, 13, 15, 19, 22] have considered this step.foraging is making

- *Remote execution control*. The final step involves the establishment of a reliable connection between the mobile device and the appropriate surrogate to pass required information, remote execution, and the receipt of returned results. Various researches [3, 4, 6, 12, 19]have considered remote execution control.

In this paper, we focus on *scheduling* step of cyber foraging and propose a decision-making mechanism to select the best location to run a mobile device's task according to the pre-gathered context information.

3.3. Cyber Foraging Goals

Cyber foraging is a solution to execute resource intensive applications on resource constrained mobile devices. In fact, available researches in cyber foraging have tried to augment some resources of mobile devices in terms of effective metrics to achieve more efficient application execution. The most important resources have been considered by offloading approachesare as follows:

- *Energy.*One of the most important constraints of mobile devices is energy consumption because mobile device's energy cannot be replenished by itself[23]. Many researches [3, 20-22] have considered energy consumption as a parameter for offloading

- *Memory and storage*. Memory capacity of mobile devices is less than stationary computers and memory intensive applications cannot usually run on mobile devices. Many researches [13-15] have considered the availability of memory and storage as another effective parameter for offloading decision.

- *Response time*. When the processing power of mobile devices is considerably lower than static computers,task offloading is beneficial to decrease execution time. There are many researches [3, 12, 14, 15, 19]that have considered the response time and latency as a major parameter affecting the offloading decision.

- *I/O*. Displaying a movie on a bigger screen, playing music on more powerful speakers, and printingare examples of task offloading to improve I/O quality or exploit more I/O devices. Some researches [16, 17] have focused on augmenting I/O as an effective parameter for offloading decision.

In this paper, we focus on energy, response time and memory. Weoffload the mobile device's tasks to decrease energy consumption and response time in mobile devices. Furthermore we consider memory demand of the task and available memory of every location (i.e. the mobile device and surrogates) to select appropriate location to execute the task.

4. PROPOSED DECISION MECHANISM

In this section we propose an approach to raise the participation rate of mobile devices in pervasive and mobile computing using context aware task offloading. The pervasive computing environment in our experiments includes a mobile device and some desktop computers as surrogates that are intra-connected through a wireless LAN.

When a task is requested to run on the mobile device, a solver program runs immediately to make a decision according to the context metrics either to offload whole the task or to execute it on mobile device itself.

4.1. Context Metrics

Due to the dynamic nature of resources involved in typical computational pervasive environments and portability of mobile devices, the ability of a device to perform the operations varies over time. Therefore, making decision to offload a task must be according to the current situation. We categorize the context metrics into four classes:
- *Mobile device metrics* include current processing power, available memory, and available energy.

- *Surrogate metrics* include current processing power, and available memory.

- *Network metrics* include network type and its current conditions that can change depending on the location such as data transmission rate and signal strength.

- *Application metrics* include application type, which is one of CPU intensive, memory intensive, and I/O intensive [25], and the size of application's code, input and output data.Because application code and input areavailable before execution, theirsize can be specified easily. Althoughoutput size is not available before task execution,in most cases it is a constant value with a knownsize or it can be estimated in terms of input valueor input size.

4.2. Solver

If we suppose to offload either the whole task or nothing and every time we make decision for only one task, we can define the solver as a formal cost function. The cost function is calculated for the mobile device and every surrogate; either the mobile device or a surrogate with minimum cost value, would be the execution platform of the task.

In this paper, we suppose having context metrics to estimate the execution cost. We define the current processing power as Equation 1, where P_u is the percentage of usage of processor, and P_s is the processor speed.

$$P_c = (1 - P_u) \times P_s \quad (1)$$

The cost function to determine the target execution location is defined by Equation 2.

$$Cost = \frac{w_1 * Time + w_2 * Energy}{w_3 * P_c + w_4 * Available\ Memory} (2)$$

Where w_1 to w_4 are the weighting factors which are non-negative values; the summation of them is one and represents the importance of the corresponding factors. Calculating the *Cost* for the mobile device, the *Time* factor is the execution time of the task on mobile device (*Time$_{mobile}$*) and the *Energy* factor referred to energy consumption at run-time that is defined as Equation 3.

$$Energy_{mobile} = Time_{mobile} \times Power_{Comp} \quad (3)$$

Energy consumption of mobile device in various states is different [20, 26]. Therefore, we have defined $Power_{comp}$ as power rate for computation on the mobile device. While $Power_{standby}$ is defined as power rate of mobile device on remote execution, and $Power_{send}$ and $Power_{receive}$ are power rate for sending and receiving data.

Calculating the *Cost* for surrogates, the *Time* factor is calculated by Equation 4, and the *Energy* factor is calculated by Equation 6.

$$Time_{surrogate} = Time_{send} + ExecutionTime_{surrogate} + Time_{receive} \quad (4)$$

$Time_{send}$ and $Time_{receive}$ are calculated in terms of *Transmission Data Size*, which includes the sizes of code, input data, and output data as given in Equation 5.

$$Time_{Send/receive} = \frac{TransmissionDataSize}{DataTransmissionRate} \quad (5)$$

$$Energy_{surrogate} = (Time_{send} * Power_{Send}) + (ExecutionTime_{surrogate} * Power_{Standby}) \\ + (Time_{receive} * Power_{receive}) \quad (6)$$

Figure 1 shows the pseudo code of our proposed solver.

```
Proposed_Solver()
{
  if ((Available_Memory_Mobile<Required_Memory_Application) or
(Available_Energy_Mobile<Required_Energy_Application))
  {
Mobile_In_Competition = FALSE;
  }
foreach surrogate
  {
    Calculate time and energy to offload the task();
    if ((Available_Memory_Surrogate[i]<Required_Memory_Application) or
        (Available_Energy_Mobile<Required_Energy_Surrogate[i]))
    {
      Surrogate[i].In_Competition = FALSE;
    }
}
  if (forall Surrogates: Surrogate[i].In_Competition == FALSE)
  {
    if (Mobile_In_Competition == TRUE)
LocalExecution();
else
DoNothing();
  }
else
  {
foreach Surrogate/Mobile: ifIn_Competition == TRUE
{
CalculateCost;
  }
    Execute the task on the Surrogate/Mobile with minimal Cost();
  }
}
```

Figure 1.Solver algorithm

5. EVALUATION

To quantify the effectiveness of our proposed approach, we constructed a test bed consisting of one mobile device and one surrogate whose specifications are given in Table 1. The mobile device was connected to surrogates via 802.11b/g WLAN. The context information of mobile device, surrogate, and applications were presented in XML file format. Figure 2 shows the context information of mobile device, and Figure 3 shows the context information of the available surrogate in the chosen test bed.

Table 1.Configuration of devices used in our experimentations.

Type	Processor	Memory	Operating System
Mobile	Qualcomm MSM7225™ 528 MHz	256 MB	Windows Mobile 6.5 Professional
Surrogate	Intel Core 2Duo 2.5 GHz	4 GB	Windows 7 Professional

```
<MobileDevice>
<NodeContext>
<Name> Mobile </Name>
<CPU> 524MHz </CPU>
<InstructionPSecond> 270 </InstructionPSecond>
<Load> 0.05 </Load>
<AvailableMemory> 91MB </AvailableMemory>
<AvailableBattery> 800J </AvailableBattery>
</NodeContext>
</MobileDevice>
```

Figure 2.Context specification of the mobile device

```
<Surrogates>
<NodeContext>
<Name> Surrogate1 </Name>
<CPU> 5000MHz </CPU>
<InstructionPSecond> 938010 </InstructionPSecond>
<Load> 0.1 </Load>
<AvailableMemory> 2200MB </AvailableMemory>
<Bandwidth> 1KB/S </Bandwidth>
</NodeContext>
</Surrogates>
```

Figure 3.Context specification of the surrogates

We evaluated the effectiveness of our proposed approach with respect to responsiveness and resource consumption. We evaluated the responsiveness of the proposed approach through a scenario where the user intended to execute an application for finding the nth prime number, which is a CPU intensive application, and needed high computing power and low memory size on a mobile device where a surrogate was in range. Figure 4 shows the context information of the nth prime application.

We evaluated the proposed approach with respect to resource consumption through a scenario where the user intended to execute an application to determine a matrix determinate, which needed high computing power and size of input data was respectively high. Figure 5 shows the context information of matrix determinate application.

As we have stated earlier, in this paper we suppose context descriptor files for the mobile device, surrogates and tasks are prepared in advance. Furthermore, in both mentioned benched applications, output data has a constant size which is indicated by *BaseOutputSize* tag in Figure 4 and Figure 5. We measured the response time and resource consumption in three scenarios: local execution of application on mobile device, offloading the application and execution on surrogate, and using our proposed method to find the target execution location and run it.

```
<ApplicationContext>
<Name> Nth Prime Number </Name>
<RequiredMemory> 0.6MB </RequiredMemory>
<CodeSize> 1KB </CodeSize
<BaseInputSize> 0.05KB </BaseInputSize>
<BaseOutputSize> 0.05KB </BaseOutputSiz>
<Order>
    (N*ln(N)+(N*ln(ln(N))))*(pow(N*ln(N)+(N*ln(ln(N))),0.5))
</Order>
</ApplicationContext>
```

Figure 4.Context specification of the *n*th prime number application

```
<ApplicationContext>
<Name> Matrix Determinant </Name>
<RequiredMemory> 9MB </RequiredMemory>
<CodeSize> 2KB </CodeSize
<BaseInputSize> 0.1KB </BaseInputSize>
<BaseOutputSize> 0.05KB </BaseOutputSiz>
<Order> N! </Order>
</ApplicationContext>
```

Figure 5.Context specification of the matrix determinant

5.1. Responsiveness

Responsiveness is defined as the time used by an application to respond to a user-triggered request [21]. In general, lower response time is more satisfactory and always we hope that response time is low enough for good subjective performance.

To estimate the execution time, we replace variable N with input value of the application in the function presented in the *Order* section of the application context description. The result is then divided by *InstructionPSecond* presented in context descriptor of mobile device and surrogate to estimate the execution time of the application on mobile device and surrogate.

Figure 6 shows the response time of the *n*th prime application among increases in input size. As it is shown, our proposed approach almost always yields the least response time and thus the best location to run the application.

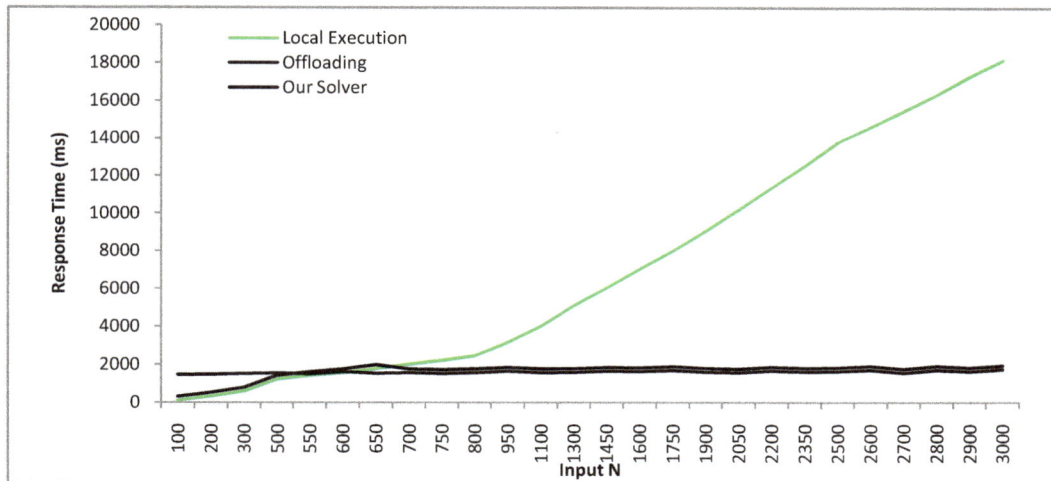

Figure 6.Comparison of execution time for finding the nth prime number

5.2. Energy

Battery lifetime is an important aspect of participating mobile devices in pervasive environments. Therefore, a good offloading mechanism should focus on consuming as low energy as possible. To evaluate the impact of cyber foraging on energy consumption, we experimentally measured the energy consumption of mobile devices through execution of a *matrix determinant* application.

In the nth prime application, a number with fixed size was the application's input data, but the *matrix determinant* application required to send the whole matrix to the surrogate. Therefore, the *Data Size* factor in Equation 5 was variable in terms of matrix's row count that affected the cost function and so the decision. In this scenario, due to simplicity, we assumed the $Power_{send}$, $Power_{receive}$, and $Power_{standby}$ factors as equal in Equation 6.

To emphasis on energy consumption in this scenario, we set the *Energy* weight (w_2) in Equation 2 to maximum value of 1. Figure 7 presents the energy consumption of execution of *matrix determinate* application; as it is shown, our proposed approach preserved the minimum energy consumption compared to local execution of the application in mobile device or always offloading the application and execution on the surrogate.

An issue that should be considered in every decision maker's mechanism is the execution overheads of the decision-making process itself, which must be as light as possible. As it is shown in Figure 6 and Figure 7, our proposed approach, preserved nearly the same response time and energy consumption compared to blind offloading approach, when it decides to offload the task; and nearly the same response time and energy consumption compared to local execution on mobile device, when it decides to execute the task on the mobile phone. Actually, the computational complexity of our proposed solver is O(n) which n is the number of available surrogates. Since the number of surrogates is always relatively small, the overhead of decision-making of our proposed solver does not affect the results and is almost negligible.

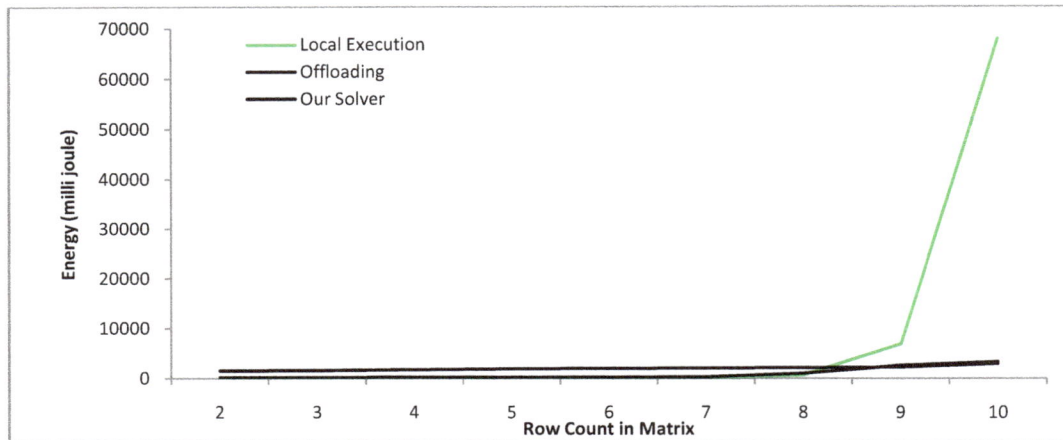

Figure 7.Comparison of energy consumption for *matrix determinant* application

6. CONCLUSION AND FUTURE WORKS

Mobile devices have always suffered from resource constraints, in comparison with static computers, to run complex and high computational applications. One of the major and most common solutions to improve computational resource poverty of mobile devices, especially in pervasive computing environments, is cyber foraging, which is offloading some tasks to more powerful nearby static computers. However, as discussed in this paper, cyber foraging is not effective in all circumstances and metrics such as mobile device and surrogate specifications, network quality, transmission data size, and application nature should be taken into account.

In this paper, we proposed a context-aware cyber foraging approach to ameliorate the resource poverty shortages of mobile devices and to raise the ability of participation of mobile devices in pervasive and mobile computing.

Experimental results showed the superiority of the proposed approach in response time and energy consumption, which are two most important metrics in mobile computing, in contrast to local execution of applications on mobile devices or blind offloading to surrogates.

As a future work, we are working on support of more than one surrogate and considering other metrics like surrogates' load, and surrogates' geographical distance that affects the wireless signal strength and network bandwidth. In addition, more experiments with various application types, and mobile/surrogates context, could increase the applicability of our proposed approach.

REFERENCES

[1]   Perry,Mark, O'hara,Kenton, Sellen,Abigail, Brown,Barry&Harper,Richard, (2001) "Dealing with Mobility: Understanding Access Anytime, Anywhere," *ACM Transactions on Computer-Human Interaction (TOCHI)*, Vol. 8, No. 4, pp. 323-347.

[2]   Balan, Rajesh Krishna, Gergle, Darren, Satyanarayanan, Mahadev&Herbsleb, James, (2007) "Simplifying Cyber Foraging for Mobile Devices," in *5th USENIX International Conference on Mobile Systems, Applications and Services (MobiSys)*, San Juan, Puerto Rico, pp. 272-285.

[3]   Flinn, Jason, Park, SoYoung&Satyanarayanan, Mahadev, (2002) "Balancing Performance, Energy, and Quality in Pervasive Computing," in *22nd International Conference on Distributed Computing Systems (ICDCS'02)*, Vienna, Austria, pp. 217-226.

[4]   Murarasu, Alin Florindor&Magedanz, Thomas,(2009) "Mobile Middleware Solution for Automatic Reconfiguration of Applications," in *6th International IEEE Conference on Information Technology*, Las Vegas, USA, pp. 1049-1055.

[5]   Chun, Byung-Gon&Maniatis, Petros,(2009) "Augmented Smartphone Applications through Clone Cloud Execution," in *12th Workshop on Hot Topics in Operating Systems (HotOS)*, Monte Verita, Switzerland.

[6]   Satyanarayanan, Mahadev, Bahl, Paramvir, Cáceres, Ramón&Davies, Nigel, (2009) "The Case for VM-Based Cloudlets in Mobile Computing," *IEEE Pervasive Computing,* Vol. 8, No. 4,pp. 14-23.

[7]   Oh, Jehwan, Lee, Seunghwa&Lee, Eunseok,(2006) "An Adaptive Mobile System Using Mobile Grid Computing in Wireless Network," in *International Conference on Computational Science and Its Applications (ICCSA 2006)*, Glasgow,UK, pp. 49-57.

[8]   Kolos-Mazuryk,L., Wieringa,R.&Van Eck,P., (2005) "Development of a Requirements Engineering Method for Pervasive Services," in *RE '05 Doctoral Consortium*, Paris,France.

[9]   M. Weiser, (1991) "The Computer for the 21st Century," *Scientific American Special Issue on Communications, Computers, and Networks,* pp. 94-104.

[10]   M. Satyanarayanan, (2001) "Pervasive Computing: Vision and Challenges," *IEEE Personal Communication,* Vol. 8, No. 4, pp. 10-17.

[11]   Balan, Rajesh Krishna, Flinn, Jason, Satyanarayanan, Mahadev, Sinnamohideen,S.&Yang, H.I., (2002) "The Case for Cybef Foraging," presented at the 10th Workshop on ACM SIGOPS European Workshop: beyond the PC, New York, NY, USA, pp. 87-92.

[12]   Balan, Rajesh Krishna, Satyanarayanan, Mahadev, Park, SoYoung&Okoshi, Tadashi,(2003) "Tactics-Based Remote Execution for Mobile Computing," in *1st International Conference on Mobile Systems, Applications and Services*, San Francisco, pp. 273-286.

[13]   Gu, Xiaohui, Messer, Alan, Greenbergx, Ira, Milojicic, Dejan&Nahrstedt, Klara, (2004) "Adaptive Offloading for Pervasive Computing," *IEEE Pervasive Computing Magazine,* Vol. 3, No. 3, pp. 66-73.

[14]   Ou, Shumao, Yang, Kun&Liotta, Antonio,(2006) "An Adaptive Multi-Constraint Partitioning Algorithm for Offloading in Pervasive Systems," in *4th Annual IEEE International Conference on Pervasive Computing and Communications (PERCOM'06)*, Pisa, Italy, pp. 116-125.

[15]   Ou, Shumao, Yang, Kun&Zhang, Qingfu, (2006) "An Efficient Runtime Offloading Approach for Pervasive Services," in *IEEE Wireless Communications & Networking Conference (WCNC2006)*, Las Vegas, pp. 2229-2234.

[16]   Song, Xiang(2008) "Seamless Mobility in Ubiquitous Computing Environments," PhD Thesis, Georgia Institute of Technology.

[17]   Song, Xiang&Ramachandran, Umakishore, (2007) "MobiGo: A Middleware for Seamless Mobility," in *13th IEEE International Conference on Embedded and Real-Time Computing Systems and Applications (RTCSA'2007)*, pp. 249-256

[18]   Kristensen, Mads Darø,(2010) "Empowering Mobile Devices through Cyber Foraging:The Development of Scavenger, an Open Mobile Cyber Foraging System," PhD Thesis, Department of Computer Science, Aarhus University, Denmark.

[19]   Kristensen,Mads Darø &Bouvin,Niels Olof, (2010) "Scheduling and Development Support in the Scavenger Cyber Foraging System," *Pervasive and Mobile Computing,* Vol. 1, No. 6, pp. 677-692.

[20]   Othrnan, Mazliza&Hailes, Stephen, (1998) "Power Conservation Strategy for Mobile Computers Using load sharing," *Mobile Computing and Communications Review,* Vol. 2, No. 1, pp. 19-26.

[21]   Kemp, Roelof, Palmer, Nicholas, Kielmann, Thilo, Seinstra, Frank, Drost, Niels, Maassen, Jason&Bal, Henri, (2009) "eyeDentify: Multimedia Cyber Foraging from a Smartphone," in *IEEE International Symposium on Multimedia (ISM2009)*, San Diego, pp. 392-399.

[22]   Cuervo, Eduardo, Balasubramanian, Aruna, Cho, Dae-ki, Wolman, Alec, Saroiu, Stefan, Chandra, Ranveer&Bahl, Paramvir, (2010) "MAUI: Making Smartphones Last Longer with Code Offload," in *8th international conference on Mobile systems, applications, and services (ACM MobiSys'10)*, San Francisco, USA, pp. 49-62.

[23]   Satyanarayanan, Mahadev, (2005) "Avoiding Dead Batteries," *IEEE Pervasive Computing,* Vol. 4, No. 1, pp. 2-3.

[24]   Chun, Byung-Gon&Maniatis, Petros, (2010) "Dynamically Partitioning Applications between Weak Devices and Clouds," in *1st ACM Workshop on Mobile Cloud Computing and Services (MCS 2010)*, San Francisco, pp. 1-5.

[25]   Zhang, Jian&Figueiredo,Renato J., (2006) "Application Classification through Monitoring and Learning of Resource Consumption Patterns," in *20th IEEE International Parallel and Distributed Processing Symposium (IPDPS).*

[26]   Park, Eunjeong, Shin, Heonshik&Kim, Seung Jo,(2007) "Selective Grid Access for Energy-Aware Mobile Computing," *Lecture Notes in Computer Science(LNCS),* Vol. 4611, pp. 798-807.

14

Distance Based Energy Efficient Selection Of Nodes To Cluster Head In Homogeneous Wireless Sensor Networks

[1]S. Taruna, [2]Sheena Kohli, [3]G.N.Purohit

Computer Science Department,
Banasthali University, Rajasthan, India
[1]staruna71@yahoo.com
[2]sheena7kohli@gmail.com
[3]gn_purohitjaipur@yahoo.co.in

ABSTRACT

Wireless sensor networks (WSN) provide the availability of small and low-cost sensor nodes with capability of detecting, observing and monitoring the environment, along with data processing and communication. These sensor nodes have limited transmission range, processing and storage capabilities as well as their energy resources. Routing protocols for wireless sensor networks are responsible for maintaining the energy efficient paths in the network and have to ensure extended network lifetime. In this paper, we propose and analyze a new approach of cluster head selection by a homogeneous sensor node (having same initial energy) in wireless sensor network, which involves choosing the cluster head which lies closest to the midpoint of the base station and the sensor node. Our proposed routing algorithm is related with energy and distance factors of each nodes. This scheme is then compared with the traditional LEACH protocol which involves selecting the cluster head which is nearest to the particular node. We conclude that the proposed protocol effectively extends the network lifetime with less consumption of energy in the network.

KEYWORDS

Wireless Sensor network, Cluster head, Routing protocol, Network lifetime, Dead nodes.

1. INTRODUCTION

A wireless sensor network (WSN) consists of numerous small autonomous devices called sensors or nodes, capable of sensing the environment, processing the information locally and sending it to the point of collection through wireless links in a particular geographical area. WSNs are scalable and smart. The sensors can communicate directly among themselves or to some base station deployed externally in the area. But being autonomous nodes , they have limited battery , processing power and bandwidth. Of all the resources constraints, limited energy is most concerning one . One of the main design goals of WSNs is to carry out energy efficient data communication while trying to prolong the lifetime of the network.[6]

Routing in wireless sensor networks is very challenging due to the essential characteristics that distinguish wireless sensor networks from other wireless networks. It is highly desirable to find the method for energy efficient route discovery and relaying of data from sensor node to base station so that lifetime of network is maximized.

Much research has been done in recent years and still there are many design options open for improvement. Thus, there is a need of a new protocol scheme, which enables more efficient use of energy at individual sensor nodes to enhance the network survivability.[5]

In this paper, we analyze energy efficient homogeneous clustering head selection algorithm by a sensor node for WSN. We first describe the new distance based scheme, its pseudocode, flowchart, packet format, and its different scenarios, and then the simulation results in MATLAB[2].

Further, the performance analysis of the proposed scheme is compared with benchmark clustering algorithm LEACH[4].

2. RELATED WORK

Routing is a process of selecting a path in the network from source to destination along which the data can be transmitted. Various protocols [3] like LEACH, HEED, PEGASIS, TEEN, APTEEN are available to route the data from node to base station in WSN.

Sensors organize themselves into clusters and each cluster has a leader called as cluster head(CH), i.e. sensor nodes form clusters where the low energy nodes called cluster members (CM) are used to perform the sensing in the proximity of the phenomenon. For the cluster based wireless sensor network, the cluster information and cluster head selection are the basic issues. The cluster head coordinates the communication among the cluster members and manages their data.[1]. The process of clustering in routing provides an efficient method for maximizing the lifetime of a wireless sensor network by rotating the role of cluster head.

Low-energy adaptive clustering hierarchy (LEACH)[4] is a popular energy-efficient clustering algorithm for sensor networks. LEACH randomly selects a few sensor nodes as CHs and rotate this role to evenly load among the sensors in the network in each round. In LEACH, the cluster head (CH) nodes compress data arriving from nodes that belong to the respective cluster, and send an aggregated packet to the base station. A predetermined fraction of nodes, p, elect themselves as CHs in the following manner. A sensor node chooses a random number, r, between 0 and 1. If this random number is less than a threshold value, T(n), the node becomes a cluster-head for the current round. The threshold value is calculated based on an equation that incorporates the desired percentage to become a cluster-head, the current round, and the set of nodes that have not been selected as a cluster-head in the last (1/p) rounds, denoted by G. It is given by:

$$T(n) = \frac{p}{1 - p(r \bmod (1/p))} \quad if \ n \in G$$

Here, G denotes the set of nodes involved in the selection of CH. Each elected CH broadcasts a message to the rest of the nodes in the network to inform that it is the new cluster-head. A sensor node or non- CH selects the CHs which is nearest to it .

LEACH clustering terminates in a finite number of iteration, but does not guarantee good cluster head distribution. Some nodes may choose a cluster so that the distance between its CH and sink (base station) is even further than the distance between the node itself and the sink. According to the energy model of LEACH protocol, the energy cost will increase as the distance increases. Battery power being limited in the sensor nodes, let the nodes to expire on full consumption of energy.

3.THE PROPOSED ALGORITHM

In order to save the total energy cost of the sensor networks and prolong its lifetime, we propose a distance-based clustering protocol, LEACH-MP (LEACH-minimal path). The basic idea of the protocol is as follows:

Firstly some assumptions are addressed in this paper:

- All nodes can send data to Base station (BS).
- The BS has the information about the location of each node. It's assumed that the cluster heads and nodes have the knowledge of its location.
- Data compression is done by the Cluster Head.
- In the first round, each node has a probability p of becoming the cluster head.
- All nodes are of same specification.
- All nodes in the network are having the same energy at starting point and having maximum energy.
- Energy of transmission depends on the distance (source to destination) and data size.
- Nodes are uniformly distributed in network in a random manner.

Like LEACH, the operation of LEACH-MP is also divided into rounds. Each round begins with a set-up phase and steady phase. We do not change the way LEACH elects its cluster heads but changed the cluster formation algorithm. After the cluster heads are selected, cluster-heads broadcast an advertisement message that includes their node ID as the cluster-head ID and location information to inform non-cluster head nodes. Non-cluster head nodes first record all the information from cluster heads within their communication range. Then the node finds the cluster head which is closest to the middle-point between the node itself and the sink and joins that cluster. In other words, we changed the way how nodes join the cluster in order to decrease the total energy cost of the network and prolong the network lifetime.

The round diagram of LEACH-MP can be seen in the following Figure 1.

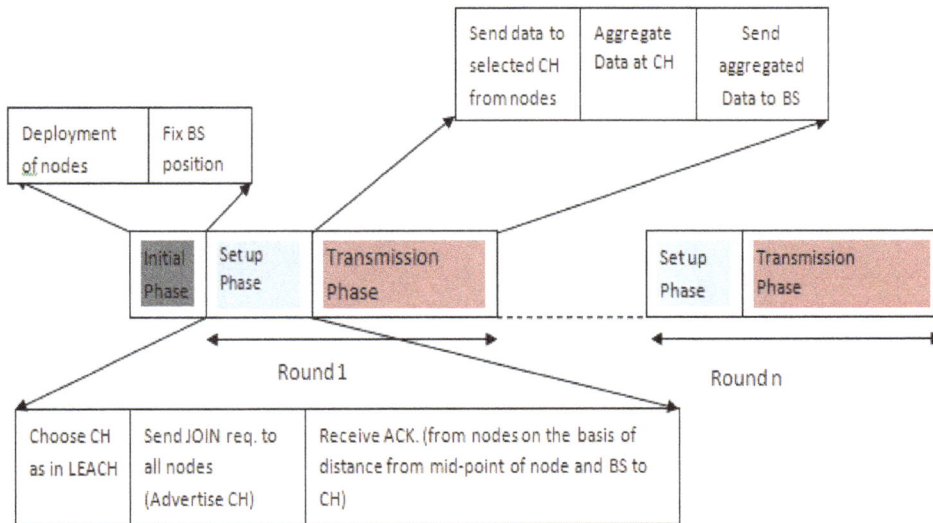

Figure.1. Round Diagram of LEACH-MP

Next the mathematical analysis will be given about the new scheme.

4. NETWORK AND ENERGY CONSUMPTION MODEL

The transmission energy of transmitting a k-bit message over a distance t is given by:

$$E_{TX(k,t)} = E_{TX\text{-}elec(k)} + E_{TX\text{-}amp(k,t)}$$
$$= kE_{elec} + k\,E_{fs}\,t^2 \tag{1}$$

E_{elec} is the transmitter circuitry dissipation per bit , E_{amp} is the transmit amplifier dissipation per bit and E_{fs} is the dissipation energy per bit.

The receiving energy cost is:

$$E_{RX(k)} = E_{RX-elec(k)}$$
$$= kE_{elec} \qquad (2)$$

The total energy cost of a network is given by:

$$E_{total} = E_{TX} + E_{RX} + E_I + E_S \qquad (3)$$

which needs to be minimized i.e. Min (E_{total})

Here , E_I is the energy cost during idle state. E_S is the energy cost while sensing.

Generally the three cost except the transmission cost are constant for a node. Only E_{TX} needs to be considered. So, we have to find Min (E_{TX}).

The energy cost mainly depends on the distance t , taking all other variable in the equation constant. Thus, we derive that we have to optimize Min (t^2). The distance between a sensor node and a cluster head is denoted as dNtoCH and that between a cluster head and a sink as dCHtosink. According to the energy model, we further simplify the optimization goal to minimize t^2 as Min (dNtoCH^2 + d CHtosink^2). As shown it the Figure 2, let the triangle SNC depicts the position of node, CH and sink or Base Station respectively, where, S is the BS , C the CH and N the sensor node. The distance between sensor node N and sink S is dNtoSink = z , dNtoCH=y and dCHtosink=x

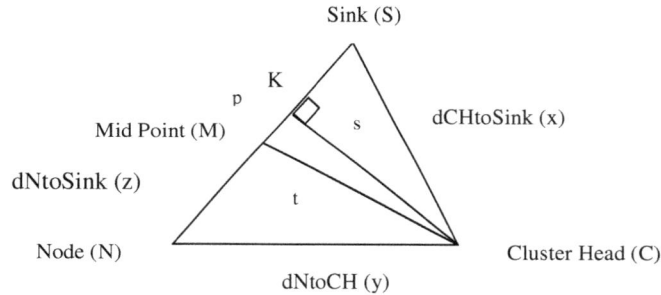

Figure 2. The basic concept

Further a perpendicular is drawn from C on line SN at point K. The length of this perpendicular is s . M is the mid point between node and sink. The distance between mid point of node and sink and that of CH is given by t.
Thus. KM=p , CM= t, CK=s

From the rule of trigonometry, applying pythagoras theorem,

In $\triangle$ SKC,
dCHtoSink^2 = CK^2 + (dNtoSink / 2 − p)^2
$$x^2 = s^2 + (z/2 - p)^2 \qquad (4)$$

In $\triangle$NKC,
dNtoCH^2 = CK^2 + (dNtoSink / 2 + p)^2
$$y^2 = s^2 + (z/2 + p)^2 \qquad (5)$$

Combining both equations (4) and (5),
$$x^2 + y^2 = s^2 + (z/2 - p)^2 + s^2 + (z/2 + p)^2$$
$$= 2s^2 + z^2/2 + 2p^2 \qquad (6)$$

From $\triangle$ MKC ,
$t^2 = p^2 + s^2$, so substituting $p^2 = t^2 - s^2$
we get , $x^2 + y^2 = z^2/2 + 2t^2$ (7)

We can see that when the value of dNtoSink is fixed , $dNtoCH^2 + d\,CHtosink^2$ is only related to t i.e. Min $(dNtoCH^2 + d\,CHtosink^2)$ is equivalent to Min (t^2). As a result , if a node chooses its CH which is closest to the mid point of this node and the sink , the squared distance of their communication is smallest. Min $(dNtoCH^2 + d\,CHtosink^2)$ is to actually minimize the distance between the CH and the midpoint of a node and the BS when the distance between the node and the BS is fixed. Thus, in LEACH–MP, non–cluster nodes select the CH which is nearest to the midpoint between itself and the BS as its communication CH for minimizing the communication cost.

The flowchart of the proposed scheme is as follows :

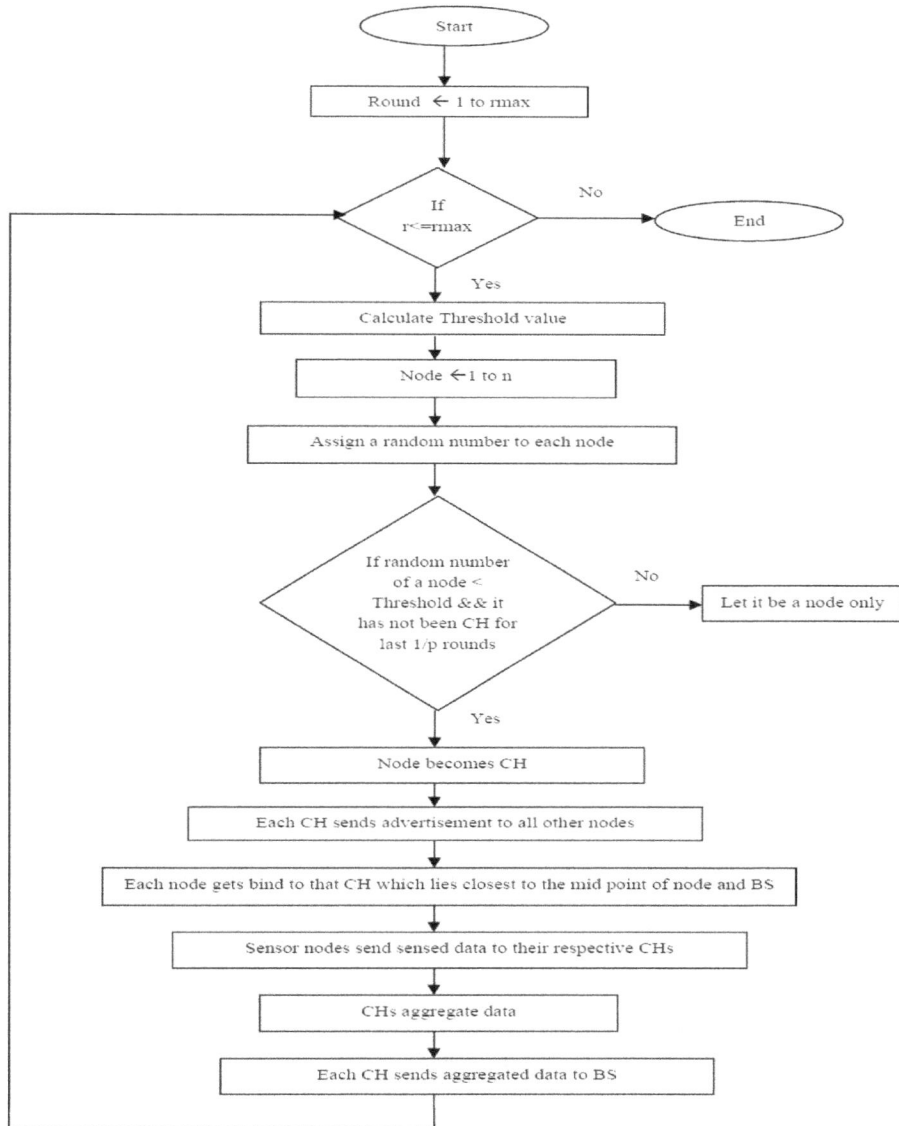

Figure 3. Flowchart of proposed scheme

The pseudocode of the proposed scheme is as follows:

1. Deployment of Sensor Nodes and Base Station in the network

The nodes are randomly deployed in the given network area of predefined size and the Base Station (BS) is located at the desired position. The nodes are shown by symbol 'o' and BS by 'x'.

2. Choosing Cluster Head

1 {For each round
2 Threshold is set to (P / (1 – P * (round % 1/P)))
3 {for each node
4 {if number of cluster head<= P && energy of node > 0
5 Assign a random number
6 { if (random number < threshold value) && (the node has not been cluster head)
7 Node is Cluster head //assign node id to cluster head list
8 Increment cluster head count //a new cluster head has been added
9 Else go to the next node}
10 Else go to the next node}
11 }

After the cluster heads are chosen for a round, clusters are ready to be formed. Sensor nodes then choose a cluster head by the one that is the closest to the mid point between it and BS.

3. Forming Clusters

1 {For each node
2 if node is a cluster head
3 go to next node
4 {else
5 {for each cluster head
6{ node coordinate x is assigned to x1
7 node coordinate y is assigned to y1
8 cluster head coordinate x is assigned to x2
9 cluster head coordinate y is assigned to y2
10 mid point coordinate x of node coordinate and BS coordinate is assigned to x3
11 mid point coordinate y of node coordinate and BS coordinate is assigned to y3
12 }
13 the distance between mid point of node to BS and cluster head is the least distance
14 cluster head id is assigned as cluster head to that node
15}
16}
17}

Transmission and Reception of Data

The cost of transmitting a message is: $ETX(k, t)=Eelec * k + Eamp * k * t^2$. The cost of receiving a message is: $ERX(k)=Eelec * k$.

18 {if distance between node and cluster head is <= the transmission range
19 Transmission cost is $E_{Tx}(k, t)=E_{elec} * k + E_{fs} * k * t^2$
20 Reception cost is $E_{Rx}(k)=Eelec * k$
21 Subtract the transmission cost from the sending node
22 {if remaining energy <= 0
23 display node has died

24 exit the program
25 }
26 Subtract the reception cost from the receiving node
27 {if remaining energy <= 0
28 display node has died
29 exit the program
30 }
31 return the sum of transmission cost and reception cost and calculate the residual energy of each node
32 }
33 }

The packet formats used in the proposed scheme will be :

1. Cluster head sends advertisement or JOIN request to all nodes in the network. The control packet comprises of following fields:

Node (CH) ID	Node type 'CH'	Location	JOIN Request =0	Blank for node ID	TTL

- Node ID or CH ID is the identifier of the node, which is a number ranging between 1 to 100 for 100 nodes.
- Node type is 'CH' if the node is cluster head else its 'N'.
- Location gives the coordinates of the position of the node in the network field.
- JOIN request is the advertisement sent by CH to other nodes to let nodes make its presence. The bit is set to '0'.
- TTL is the Time To Live, here it is 1 round . The request will be valid for each round.
- A blank field is sent with the request to all nodes for attaching the node ID , by the node which accepts the request.
- The control packet size is 500 bits.

2. Nodes choose their Cluster Head and send ACK.

Node (CH) ID	Node Type 'CH'	Location	JOIN Reply=1	Node ID	TTL

- The acknowledgement for the JOIN request is sent by setting the bit as '1' , if the node accepts the request of the cluster head to get itself bind to the CH.
- Here the bit is set to '1' and the node which sends the acceptance reply, attaches its ID in the blank field.

3. Nodes send data to that CH. The data packet comprises of following fields:

Node ID	CH ID	Data

- Node ID is the identifier of the node, which is a number ranging between 1 to 100 for 100 nodes.
- CH ID is the identifier of the CH to which the node sends the data.
- Data is the actual data that has to be transmitted from node to CH and then further to BS after compression.
- The data packet size is 4000 bits.

4. The Cluster Head sends data to BS after compression. The Compression rate or fusion rate is taken as 0.6

Node (CH) ID	BS ID	Aggregated data

- Node (CH) ID is the identifier of the node , which is a number ranging between 1 to 100 for 100 nodes.
- Aggregated data is the data obtained on compression by CH at cc of 0.6
- BS ID is the base station identifier which is needed by CH to send data to the base station , more specifically when there are more than one BS.

The tables to be maintained at each Cluster Head and Node are :

(1) At CHs

CH ID	Residual Energy	Round no. when it becomes CH	Location	Node IDs to which it sends the request	Node IDs from which it gets the acceptance	Data buffered	BS ID
----	----	----	----	----	----	----	----
-----	----	----	----	----	-----	-----	----

(2) At Nodes

Node ID	Location	Residual energy	CH IDs from which request has been obtained	CH IDs to which acceptance has been send	Data
-----	----	----	----	----	----
-----	----	----	-----	-----	----

5. SIMULATION AND PERFORMANCE EVALUATIONS

We choose MATLAB for simulation. The protocol is compared to the LEACH algorithm giving results on the comparison of energy consumption and network lifetime under different scenarios.

5.1. Simulation Parameters

Table 1. Simulation Parameters

Parameter	Values
Simulation Round	2000
Number of nodes	100
CH probability	0.1
Fusion rate (cc)	0.6
Initial node power	0.5 Joule
Nodes Distribution	Nodes are randomly uniformly distributed
Packet size (k bits)	4000
Energy dissipation (Efs)	10*0.000000000001 Joule
Energy for Transmission (E_{TX})	50*0.000000000001 Joule
Energy for Reception (E_{RX})	50*0.000000000001 Joule
Energy for Data Aggregation (EDA)	5*0.000000000001 Joule

5.2. Simulation Results

5.2.1. Energy Consumption with different sink location

The energy consumption of LEACH-MP protocol has been compared with that of LEACH by changing the location of Base station. The simulation was done on a network of area 200 x 200 and the energy consumed in the network was calculated by letting BS to be at different locations i.e. (100,100), (100,200),(100,250) and (100,300). The graphs shown in Figure 4(a) and (b) depict that in each case, the energy consumption of LEACH-MP is always less than that of LEACH, even on changing different BS locations in the network .

Table 2. Energy consumption on changing BS position

Sim.Run	Energy consumption in µJ							
	BS(100,100)		BS(100,200)		BS(100,250)		BS(100,300)	
	leach MP	leach	leach MP	leach	leach MP	leach	leach MP	Leach
1	33.62	42.21	23.1	125.662	125.79	162.24	149.61	273.61
2	16.8	75.89	10.57	71.73	71.96	338.99	115.58	177.6
3	42	91.78	43.1	212.7	24.62	216.03	112.08	156.21
4	37.5	104.3	33.21	235.65	16.61	101.52	193.39	410.42
5	25.9	56.11	26.61	167.71	38.34	153.9	92.61	458.33
6	15	26.1	16.71	38.09	192.43	310.09	87.82	377.91
7	62.7	208.09	18.01	38.8	74.65	122.02	32.81	200.62
8	45.25	98.12	50.98	197.06	54.8	232.97	84.44	371.97
9	75.75	185.1	64.34	128.34	16.1	82.58	108.82	391.55
10	46.23	66.42	70.12	169.23	28.97	64.26	136.93	182.06

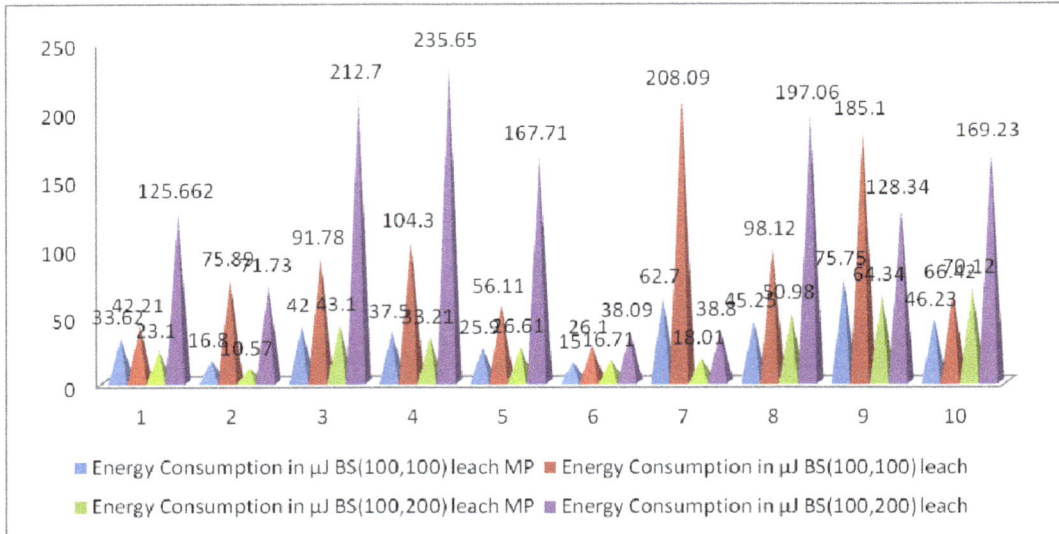

Figure 4(a). Energy consumption v/s simulation run on changing BS position

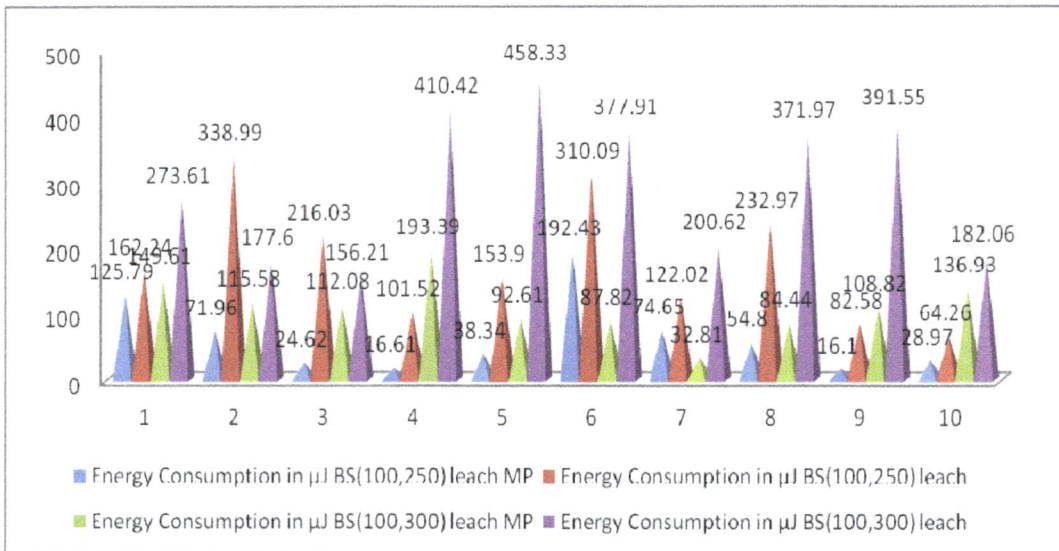

Figure 4(b). Energy consumption v/s simulation run on changing BS position

5.2.2. Network lifetime with different sink location

For analyzing the network lifetime, the number of nodes which became dead after each simulation run was compared for both the protocols. A node is declared dead when its energy becomes less than zero and hence it cannot contribute to the network any more. We can conclude that LEACH-MP extends the network lifetime as compared to LEACH, as the number of nodes which died in the end of each simulation is less in

LEACH-MP than that in LEACH, no matter where the sink is located, keeping network size as 200x200. The result is shown in Figure 5.

Table 3. Network lifetime on changing BS position

| Sim. Run | No. of Dead Nodes | | | |
| | BS(100,100) | | BS(100,200) | |
	leach	leach MP	leach	leach MP
1	58	4	81	46
2	64	10	83	49
3	54	1	89	77
4	60	0	90	74
5	57	5	79	42
6	63	4	80	48
7	41	6	75	44
8	60	5	81	50
9	41	2	73	56
10	50	3	82	51

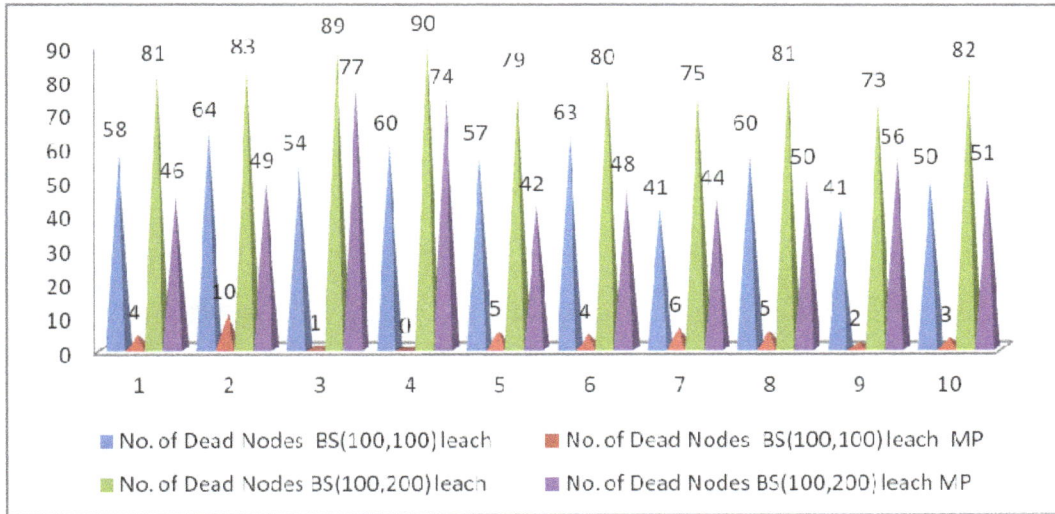

Figure 5. No. of dead nodes v/s simulation run on changing BS position

5.2.3. Energy Consumption with different network size

Further the energy consumption of the network was calculated by changing the size of the network area. i.e. (50x50), (100x100), (200x200) and (500x500). The BS is located at the centre in each case.

The simulation results are shown in Figure 6(a) and (b). The consumption of energy is more in LEACH as compared to LEACH-MP protocol in each case, even on changing the network size.

Table 4. Energy consumption on changing network size

Sim.Run	Energy consumption in µJ							
	N/w size		N/w size		N/w size		N/w size	
	50X50		100X100		200X200		500X500	
	leach MP	leach	leach MP	leach	leach MP	leach	leach MP	Leach
1	2.48	6.07	11.2	39.7	122.21	259.93	539.64	763.5
2	3.92	6.26	10.59	31.25	107.43	172.88	88.99	367.39
3	2.28	5.68	13.28	21.8	93.02	178.58	289.07	363.14
4	2.58	8.1	7.04	12.32	7.53	112.22	42.22	138.91
5	4.2	14.56	19.9	61.34	55.51	352.21	577.64	840.96
6	8.48	11.17	6.25	14.18	11.37	20.08	175.02	554.44
7	1.92	6.01	2.64	17.7	53.19	140.38	185.17	639.78
8	3.38	4.52	10.96	38.71	19.04	50.78	364.85	991.69
9	1.87	5.2	13.31	20.71	66.2	261.34	30.78	144.05
10	1.35	8.31	18.38	31.67	82.84	275.05	141.52	493.33

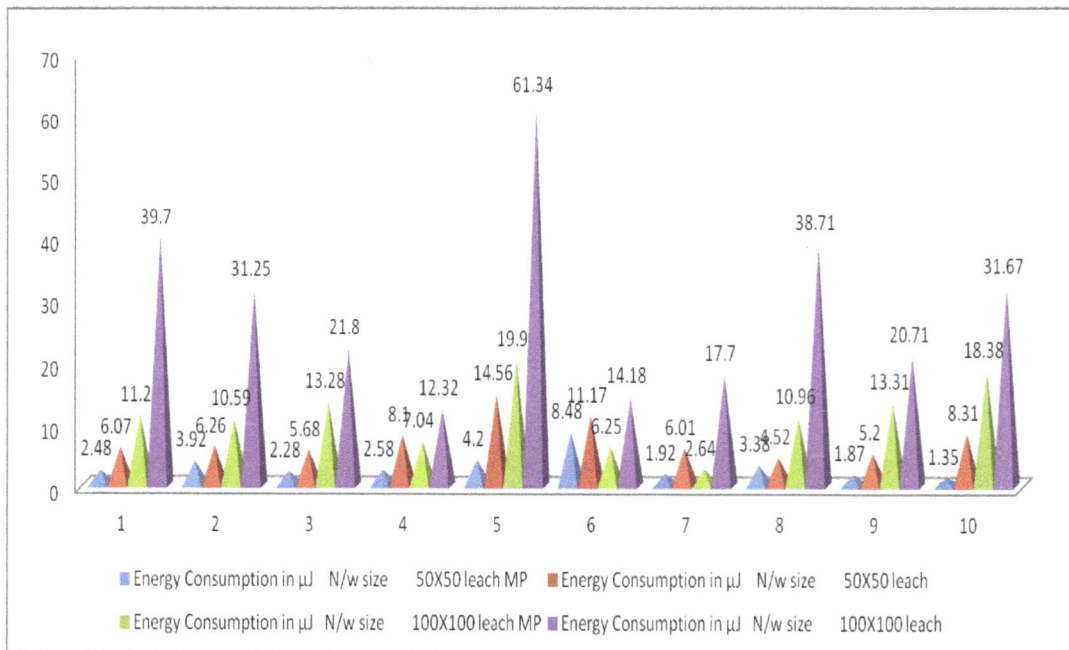

Figure 6 (a). Energy consumption v/s simulation run on changing network size

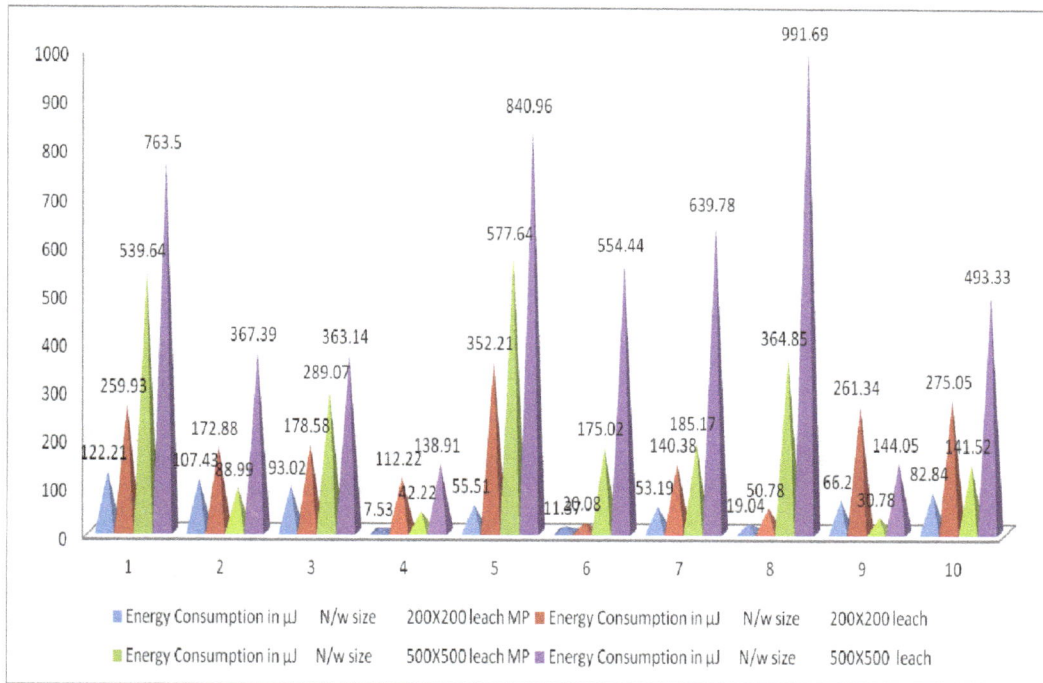

Figure 6 (b). Energy consumption v/s simulation run on changing network size

5.2.4. Network lifetime with different network size

Similarly , we changed the network area size again and calculated the lifetime of the network with both the protocols. The results are shown in Figure 7.

It was concluded that the number of nodes becoming dead at the end of each simulation run was more in LEACH-MP as compared to LEACH in each case, whatever the network size be

Table 5. Network lifetime on changing network size

Sim. Run	No. of Dead Nodes			
	N/W Size 200x200		N/W size 500x500	
	leach	leach MP	leach	leach MP
1	57	7	85	32
2	58	0	82	35
3	56	1	77	44
4	55	3	79	29
5	60	8	81	37
6	61	5	83	48
7	63	9	80	49
8	64	12	78	46
9	57	3	88	40
10	58	14	80	45

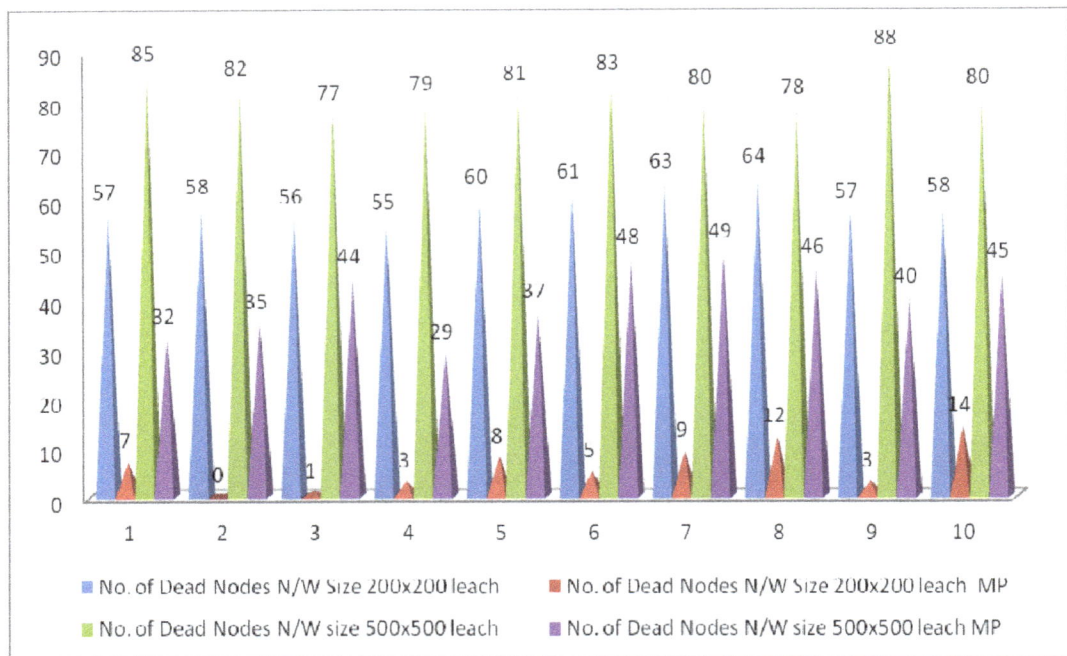

Figure 6. No. of dead nodes v/s simulation run on changing network size

6. CONCLUSION AND FUTURE WORK

Energy consumption is the main design issue in routing of Wireless Sensor Networks. We concluded that energy consumed for the cluster head selection is less in the proposed algorithm, where we choose the cluster head which lies closest to the midpoint of the base station and the sensor node , which directly shows the increased network survivability. Further the network lifetime, in terms of number of nodes dead after each simulation, of the proposed algorithm has greater span than the LEACH protocol, even on changing the network size and sink position. The proposed algorithm is for the homogeneous network and we propose to extend this work for the heterogeneous network.

REFERENCES

[1]   J. Al-Karaki, and A. Kamal, "Routing Techniques in Wireless Sensor Net-works: A Survey", IEEE Communications Magazine, vol 11, no. 6, Dec. 2004, pp. 6-28.

[2]   http://www.mathworks.in

[3]   Ankita Joshi , Lakshmi Priya.M "A Survey of Hierarchical Routing Protocols in Wirless Sensor Network", International Conference on Information Systems, May 2011, pp 67-71

[4]   Wendi Rabiner Heinzelman et al."Energy-Efficient Communication Protocol for WirelessMicrosensor Networks", in Proceeding of the 33rd Hawaii International Confrence on System Sciences, January 2000,pp1-10

[5]   S. Taruna, Jain Kusum Lata, Purohit G.N, "Zone Based Routing Protocol for Homogeneous Wireless Sensor Network", International Journal of Ad hoc, Sensor & Ubiquitous Computing (IJASUC) Vol.2, No.3, September 2011, DOI : 10.5121/ijasuc.2011.2307 99

[6]   S. Chandramathi, U.Anand, T.Ganesh, S.Sriraman and D.Velmurugan , "Energy Aware Optimal Routing for Wireless Sensor Networks", Journal of Computer Science 3 (11): 836-840, 2007, ISSN 1549-3636© 2007 Science Publications

15

COMPARATIVE EVALUATION OF FADING CHANNEL MODEL SELECTION FOR MOBILE WIRELESS TRANSMISSION SYSTEM

Z.K. Adeyemo[1] D.O. Akande, F.K. Ojo and H.O. Raji

Department of Electronic & Electrical Engineering, Ladoke Akintola University of Technology, Ogbomoso, Nigeria.
zkadeyemo@yahoo.com, akandedamilare@yahoo.com,
festusoluseye@yahoo.com, hammyraj@gmail.com
[1]Author to whom correspondence should be sent

ABSTRACT

This paper compares the performance of Rayleigh, Rician and Log-normal distributions in mobile wireless transmission system. In these distributions link quality cannot be predicted accurately enough to achieve reliable performance because of the unpredictable nature of the time-varying channel at different mobile speeds in sub-urban environment and as a result it will be very difficult to select the appropriate model in the system design for proper implementation. Binary Phase Shift Keying signaling scheme is used to modulate the data to be analyzed for efficient transmission over the three different distributions at mobile speeds of 30 km/h, 60 km/h and 90 km/h at carrier frequency of 900 MHz. Square root raised cosine filter is used to convert the modulated signal to analog signal suitable for transmission. The received bits through the channel without mitigating technique over the three scenarios are analyzed and Bit Error Rate (BER) computed for comparison. The results obtained showed that Rayleigh distribution has the highest BER, while Rician distribution has the BER between the Rayleigh and log-normal at the different speeds considered. This implies that Rayleigh distribution is the best model to be adopted by communication systems engineers.

KEYWORDS

Multipath, BPSK, Bit Error rate, Rayleigh, Rician, Log-normal

1. INTRODUCTION

Wireless communication is one of the most vibrant areas in the communication field today due to a confluence of several factors. In telecommunication, wireless communication may be used to transfer information over short and long distances. It encompasses various types of systems like fixed and mobile portable two-way radios, cellular telephones, personal digital assistants (PDAs), and wireless networking [1]. Wireless operations permits services, such as long range communications, that are impossible or impractical to implement with the use of wires. The transmission services offered by wireless communication can be categorized into fixed wireless transmission and Mobile wireless transmission. In fixed wireless transmission, the transmitter and the receiver are static while mobile wireless transmission, the transmitter and the receiver are in motion but are moving at different speeds [2,3,4,5].

Channel is the physical medium that is used to send the signal from the transmitter to the receiver [6]. More recently, it is also used to describe a radio terminal that is attached to a high speed mobile platform. According to [6], there are several kinds of communication impairments that are typical of the mobile wireless environment. Impairments may result mainly from multipath transmission, attenuation of signal power from large objects, relative transmitter-receiver motion, interference, spreading of electromagnetic power and thermal or background noise. Multipath propagation in connection with the movement of the receiver and/or the transmitter leads to drastic and random fluctuations of the received signal thus resulting into a phenomenon referred to as fading. In mobile radio communication, the emitted electromagnetic waves often do not reach the receiving antenna directly due to obstacles blocking the line-of-sight path and causes reflection, diffraction, and scattering. In fact, the received signals are a superposition of waves coming from all directions. The radio spectrum available for wireless communications is extremely scarce, while demand for mobile and personal communication is growing at a rapid space [4,7,8,9,10]. The wireless channel poses a severe challenge as a medium for reliable high-speed communication. It is not only susceptible to noise, interference and other channel impediments, but nature of these impediments change over time in unpredictable ways due to user movements causing the received signal to fluctuate or vary [3]. These defining characteristics of the mobile wireless channel can be interpreted as variations of the channel strength over time and frequency. The major effect of multipath propagation is Doppler spread which has a negative influence on the transmission characteristics of the mobile radio channel. This effect causes a frequency shift of each of the partial waves [11,12]. One of the most common methods for characterizing a fading channel is the use of a probability density function (pdf), which represents the probability density of the received signal strength. The shape of a pdf determines the performance of a wireless receiver in the presence of noise and interference [7,6]. Proper characterization of fading pdfs also impacts the design and use of mitigating techniques for a communication link [3]. The received envelope is very important because it determines the range of received signal strengths. The envelope of the received signal can determine the ultimate Shannon channel capacity of a fading wireless link since received power is proportional to the square of received envelope [13].

Most parts of communication systems over time-varying channels are designed to achieve required performance under the worst-case channel conditions. Large margins are taken into account at the designed stage. A major problem is that such margins do not allow for taking maximum advantage on the available channel capacity especially in the case of varying channels [14,15,16,17,18,19,20]. Nowadays, the communication systems engineer has at their finger tips many distributions to predict the behaviour of radio communication systems over fading channel. Many researchers have worked on some of the appropriate model for different environments in fixed wireless communication. But to the best of authors' knowledge, up to now much work has not been carried out on the model selection at different mobile speeds in sub urban environment. The models/distribution to be evaluated are Rayleigh, Rician and Log-normal distributions, hoping that the one with the highest BER will be the best model to choice in the design of communication system before the implementation. The generated data is converted to bits, reshaped and modulated with BPSK signaling scheme. Square root raised cosine filter of a roll-off factor of 0.25 is used to reduce the spectral occupancy and converts the BPSK modulated signal to analog signal suitable for transmission over the time-varying channel. The filter BPSK modulated signals are transmitted through Rayleigh, Rician and log-normal fading distributions. The received signal envelope is demodulated by BPSK demodulator, filtered again to obtain the faded bits. Then, BER which is used as a figure of merit for the evaluation of each of the three scenarios is obtained for comparison.

The results obtained showed that Rayleigh distribution has the highest BER at all the signal to noise ratio, while Rician distribution has its BER values between the Rayleigh and log-normal at mobile speeds of 30 km/h, 60 km/h and 90 km/h when measure at the same distance. This implies that Rayleigh fading channel is the best model to be adopted by communication systems engineers because if the system engineer can design a robust system in this environment, it will be better in other environments.

2. SYSTEM MODEL

In this paper, the system model consists of the transmitter, the three different fading scenarios and the receiver. This is shown in Figure 1. The transmitter is the sub-system that takes the information signal and processes the randomly generated data prior to transmission.

The received signal, $r(t)$, is given by:

$$r(t) = m(t) * s(t) \qquad (1)$$

where $m(t)$ is the fading distribution component

s(t) is the BPSK signal

Note that: the fading distribution component $m(t)$ in the received signal $r(t)$, is a random variable.

The input data is converted into bits, reshaped and modulated with Binary Phase Shift Keying (BPSK) signaling scheme. Square-root raised cosine filter is used to reduce the spectral occupancy and converts the modulated digital signal to analog signal suitable for transmission over the radio channel.

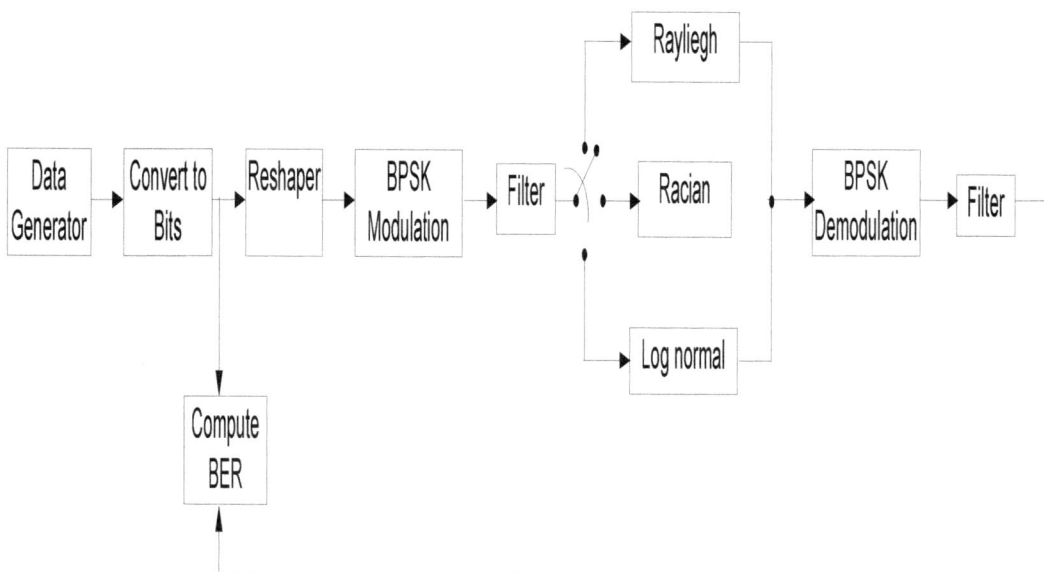

Figure 1. Fading channel selection process

2.1. Binary Phase Shift Keying (BPSK) Modulation Scheme

BPSK modulation which is a two level modulation is used for the investigation. In this paper, the phase of the carrier varies in accordance with the message signal. That is the phase of constant carrier amplitude is switched between two values according to the two possible signals corresponding to binary 1 and 0. The two phases are separated by 180^0 [21].

2.2. Analytical Expression

The BPSK signal is given as

$$S_{BPSK}(t) = \sqrt{\frac{2E_b}{T_b}} \cos 2\pi f_c t + \phi_c \qquad 0 \leq t \leq T \qquad \text{(binary 1)}$$

$$S_{BPSK}(t) = \sqrt{\frac{2E_b}{T_b}} \cos 2\pi f_c t + \phi_c + \pi \qquad 0 \leq t \leq T \qquad \text{(binary 2)} \qquad (3)$$

where E_b = energy per bit

T_b = transmitted symbol

ϕ_c = the phase

BPSK signal is generally represented as

$$S_{BPSK}(t) = m(t) \sqrt{\frac{2E_b}{T_b}} \cos 2\pi f_c t + \phi_c \qquad (4)$$

where $m(t)$ is binary data which takes on one of two possible pulse shapes.

2.3. Doppler Frequency Shift

Doppler shift is the random changes that occur in a channel introduced as a result of a mobile user's mobility or movement. It is the apparent difference in frequency of the received signals from that of the transmitted signals when there is a relative motion between the transmitter and receiver.

This Doppler frequency shift, f_d, is given by [14] as:

$$f_d = \frac{v}{c} \cos \theta \qquad (5)$$

where θ is the angle formed between the incident electromagnetic wave and the moving receiver

v is the mobile speed

λ is the wavelength of carrier.

c is speed of light.

2.4. Rayleigh Distribution

This is one of the distributions encountered in multipath propagation, this occurs when the envelope of the received signal follows a Rayleigh distribution. It is used to model locations that are heavily shadowed by surrounding buildings as Rayleigh Distribution. Rayleigh distribution is statistically used to model a faded signal when there is no dominant path.

The envelope of the received signal with Rayleigh distribution has the probability density function (pdf) given by [7] as

$$p(r) = \frac{r}{\sigma^2} \exp\left(-\frac{r^2}{2\sigma^2}\right) \qquad 0 \leq t \leq \infty \qquad (6)$$

where r is the received signal envelope.

$\frac{r^2}{2}$ is the instantaneous power.

σ is the root mean square (r.m.s) value of the received signal

σ^2 is the local average power of the received signal before envelope detection.

2.5. Rician Distribution

The Rician distribution which also occurs as a result of multipath propagation is statistically used to model a distribution when a strong line of sight component is present along with the weaker components.

It has the probability density function (pdf) given by [15] as:

$$p(r/s,\sigma) = \frac{r}{\sigma^2}\exp\left(-\frac{r^2+s^2}{2\sigma^2}\right)I_0\left(\frac{sr}{\sigma^2}\right) \qquad \text{for } s \geq 0, \quad r \geq 0 \tag{7}$$

where s is the peak amplitude of the dominant signal

$I_o(...)$ is zero order Bessel function of the first kind

$\dfrac{r^2}{2}$ is the instantaneous power

σ is the standard deviation of the local power.

Rician distribution is often described in terms of a parameter, k, known as the Rician factor and is expressed by [7] as:

$$k = 10\log\frac{s^2}{2\sigma^2} \tag{8}$$

As s approaches 0, k approaches ∞ dB and as the dominant path decreases in amplitude, the Rician distribution degenerates to a Rayleigh distribution.

2.6. Log-normal Distribution

Log-normal distribution describes the random shadowing effects as a result of buildings and other objects on the propagation path which occurs over a large number of measurement locations which have the same transmitter and receiver separation. Therefore, random fluctuations in the mean signal power occur over large distance.

The probability density function (pdf) of a log-normal distribution given by [22] is:

$$p(s) = \frac{1}{\sigma_s\sqrt{2\pi}}\exp(-\frac{(S-s_m)^2}{2\sigma_s^2}) \tag{9}$$

where s_m is the mean value of S in dBm

σ_s is the standard deviation of S in dB

S is the value of 10log(s) in dBm

s is the signal power in mW

2.7. System Performance Measure

2.7.1 Bit error rate (BER)

BER is a performance measure that specifies the number of bits destroyed as they are propagating through the distributions. In this paper, transmission speed and the distribution of the channel affect the BER performance. BER for each of the distributions is determined with equation (11).

$$\text{BER} = \frac{number\ of\ bits\ in\ error}{total\ number\ of\ bits\ sent} \tag{10}$$

The BER is given by [8] as:

$$\text{BER} = \int_0^\infty P_b\left(\frac{E}{r}\right) P(r)\, dr \tag{11}$$

where $P_b\left(\frac{E}{r}\right)$ is the conditional error probability which depends on the BPSK

modulation.

P(r) is the pdf of each of the Rayleigh, Rician, and Log-normal distributions.

2.8. Transmit and Receive Filter

In this paper, an identical square root raised cosine pulse shaping filter is used as both the transmit and receive filters so that the input signal can fit in the analog band limited channel to have zero intersymbol interference (ISI) at the pulse sampling time and to reduce the noise power outside of signal bandwidth at the receiver front end [23]. Filtering is achieved by making sure that the overall channel transfer function has a Nyguist frequency response which is the characteristic of square root raised cosine pulse shaping filter in mobile communication [8]. The impulse response of the filter $q_{rc}(t)$ is given by [8] as

$$q_{rc}(t) = \left\{ \frac{\sin\left(\dfrac{\pi t}{T_s}\right)}{\dfrac{\pi t}{T_s}} * \frac{\cos\left(\dfrac{\pi \beta t}{T_s}\right)}{1 - \left(\dfrac{4\beta t}{2T_s}\right)^2} \right. \tag{12}$$

The transfer function of a raised cosine filter P_{rc} is obtained by taking the Fourier transform of the impulse response which is given by [8] as

$$P_{rc} = \left\{ \begin{array}{ll} T_s & 0 \le |f| \le \dfrac{(1-\beta)}{2T_s} \\[3mm] \dfrac{T_s}{2}\left[1 + \cos\left(\dfrac{\pi\left(|f| \cdot 2T_s - 1 + \beta\right)}{2\beta}\right)\right] & \dfrac{(1-\beta)}{T_s} \le |f| \le \dfrac{(1+\beta)}{2T_s} \\[3mm] 0 & |f| > \dfrac{(1+\beta)}{2T_s} \end{array} \right. \tag{13}$$

where β is the roll off factor which ranges between 0 and 1

T_s is the symbol time

Since the spectrum is zero for $|f| > \dfrac{(1+\beta)}{2T_s}$, the bandwidth BW of the baseband pulse is

$\dfrac{(1+\beta)}{2T_s}$. As β increases, the bandwidth of the filter increases and the time side lobe levels decrease in adjacent symbol slots.

3. SIMULATION PARAMETERS

The investigation of the performance of the BPSK signals in each of the distributions previously mentioned is carried out using MATLAB communication and signal processing Toolboxes. The parameters that closely describe the fading channels are used. These parameters are contained in Table 1.

3.1. Acquisition of Data

The simulation process was carried out with Random data source which was converted to binary using MATLAB's decimal to binary function, reshaped and modulated with BPSK scheme for transmission to each of the fading distributions. The following parameters were used;

Table1. Simulation Parameters

Parameters	Variables
Length of message, numSymb	10000
Number of samples per symbols, nSamp	8
Modulation level, M	2
Carrier frequency of 900 MHz, f_c	900e6
Speed of light, c	3e8
Mobile speed in m/s, vel.(30, 60, 90 km/h)	(vel*1e3)/3600
Doppler spread, f_d	(2*pi* f_c *vel.)/c
Signal-to-noise ratio	0:2:14
Delay spread	$0.1 e^{-4}$ - $0.1 e^{-4}$

4. RESULTS AND DISCUSSION

The BER performance obtained from the BPSK transmission over the three different fading scenarios previously mentioned at the three different mobile speeds are presented in Figures 2, 3, and 4. These results are obtained by computer based simulations of the developed algorithms using MATLAB 7.7 application package. Figure 2 shows the simulated BER performance for Rayleigh, Rician and Log-normal distributions using BPSK modulation scheme at mobile speed of 30 km/hr. It is evident from the result that as the signal to noise ratio (SNR) increase, the BER values for the distributions also decrease. For example at SNR of 4dB, the BER value over AWGN is 0.0291, Rayleigh is 0.16533, and Rician is 0.090917 while Log-normal is 0.033597. When the SNR is increased to 8dB, the BER value for AWGN is 0.00019, Rayleigh is 0.072, Rician is 0.046917 and Log-normal is 0.010482. Also at SNR of 14dB, the values for AWGN, Rayleigh, Rician and Log-normal fading channels obtained are 0, 0.0075, 0.00625, and 0.00069669 respectively. From the Figure 2, it can be deduced that the Log-normal fading environment has the lowest BER value with the BPSK modulation technique at uniform speed of 30 km/hr. Figure 3 also shows the result obtained at mobile speed of 60 km/h for the three scenarios. It is evident from the result that as the SNR values are increasing, the BER values for the distributions are also decreasing. For example at SNR of 2dB, the BER value for AWGN channel is 0.00444, Rayleigh is 0.4045, and Rician is 0.16533 while Log-Normal is 0.033597. When the SNR is increased to 6dB, the BER value for AWGN channel is 0.0012, Rayleigh is 0.23933, Rician is 0.146 and Log-normal is 0.042001. It is observed that at mobile speed of 60km/hr, for SNR value considered BER values of the various fading environments are higher than at the mobile speed of 30 km/h. Also, log-normal environment gives the lowest BER value.

The result obtained in Figure 4 at 90 km/h for the three distributions follow the same trend except the BER values are higher at all SNR. At higher SNR, which is from 14 dB, the performance obtained through the Rayleigh and Rician are almost the same. This is because high power will be transmitted, and as a result it will be very difficult to experience the effect of dominant component in the Rician distribution. The results obtained are justifiable in that the effect Doppler spread is noticeable at high speeds and affect the BPSK transmitted signals. Also, where there is no line of sight, it means that many obstacles are along the path hence high error will occur as expected. Also since the interference levels present in a communication system are generally set by external factors due to those distributions and cannot be changed by the system design, then the bandwidth of the system are kept constant with the filter used to reduce the error to be achieved by reducing the bandwidth with the constant power level. Though, each type of modulations performs differently in the presence of different fading, therefore, BPSK signalling scheme reduces the effect of fading more than higher order of PSK modulation scheme. The highest BER values observed in Rayleigh fading indicates the highest number of error received out of 1000 bit transmitted and worst performance of the system while log-normal which gives the least bit errors received indicates the best performance with the system. The performance of thr Rician is in between. To select the best fading model in the communication design requires selecting the one which gives the highest bit errors because if the wireless communication system can perform very well in this worst model, it will perform excellently in other channels. The results obtained in this paper is in agreement with the work of [15] where many experimentally and theoretically based models have been proposed to quantitatively evaluate and combat the fading phenomenon. In his work, minimum description length (MDL) criterion for model selection is proposed and the measurements were taken in a sub-urban environment.

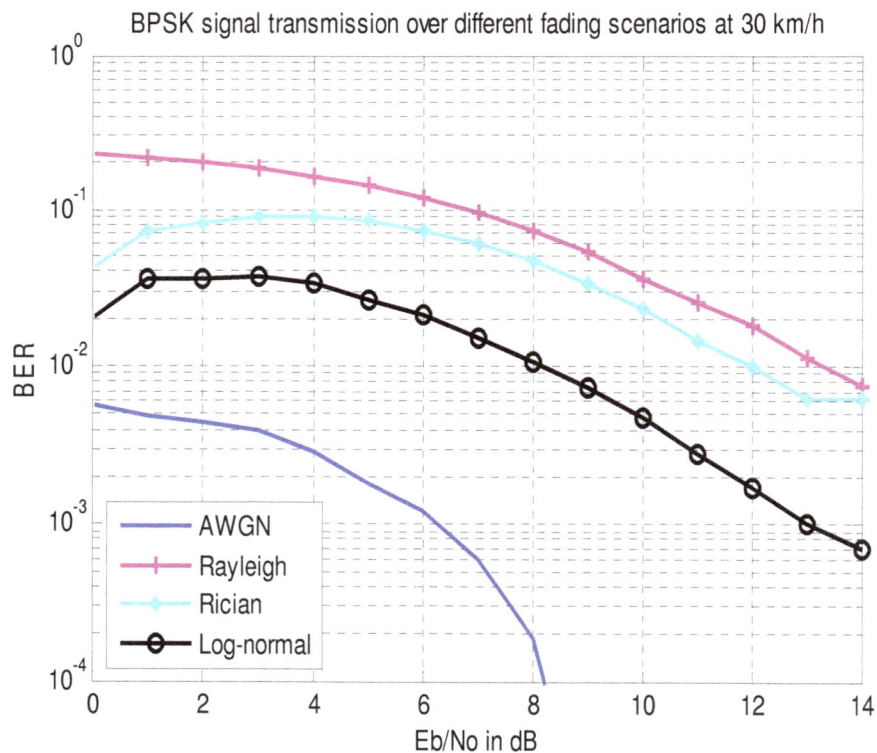

Figure 2. Simulation of BER versus SNR over Rayleigh, Rician, and log-normal distributions at 30 km/h.

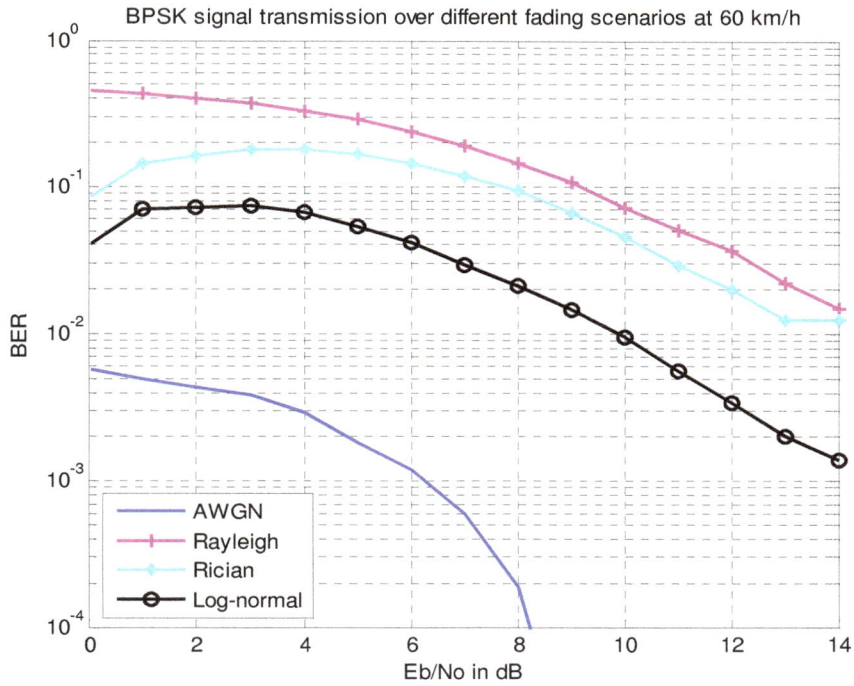

Figure 3. Simulation of BER versus SNR Rayleigh,
Rician and log-normal distributions at 60 km/h.

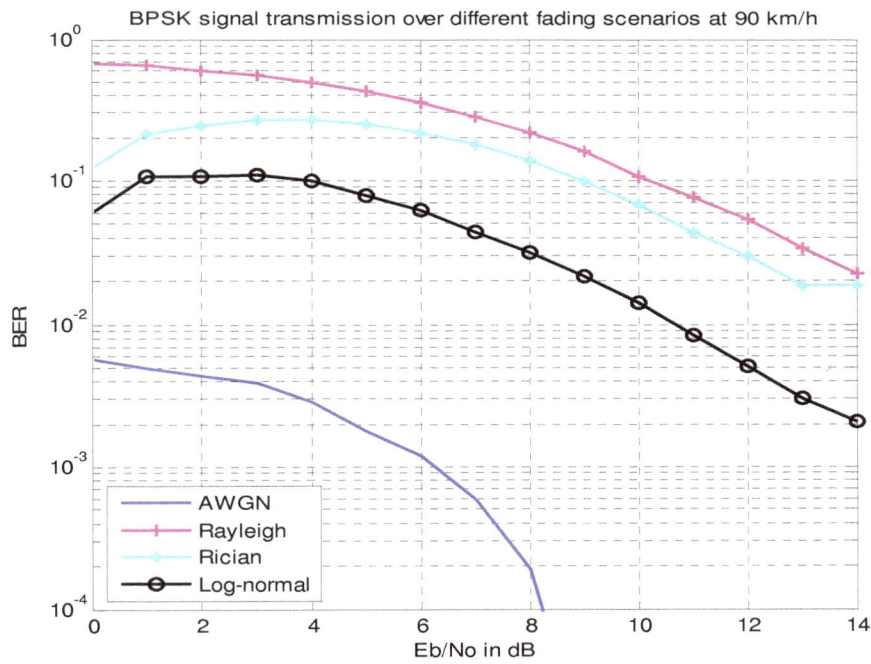

Figure 4. Simulation of BER versus SNR over Rayleigh,
Rician and log-normal distributions at 90km/hr.

5. CONCLUSIONS

In this paper, the performance of fading channel models for mobile wireless transmission system using BPSK signal with MATLAB communication Toolbox has been evaluated in term of BER. The system model for the received signal has been obtained in the presence of the Rayleigh, Rician, and log-normal distributions formed from multipath propagation and BPSK signals, the noise is modeled as the ideal (AWGN) channel. The AWGN channel served as a benchmark to determine the number of iterations to be used for the process in the sub-urban environment. BER has been obtained for each of the distributions using the probability density function of the fading distributions and the conditional error probability which depends on BPSK signal. The BER results obtained at mobile speed of 30 km/hr with the SNR of 2 dB measured at the same point for AWGN, Rayleigh, Rician and Log-normal fading are 0.00444, 0.20225, 0.082667 and 0.036196 indicating 0%, 20%, 8% and 3.2% bits error respectively. Also at an increased mobile speed of 60 km/hr with the same power, the BER values obtained for AWGN, Rayleigh, Rician and Log-normal are 0.00444, 0.4045, 0.16533 and 0.072392 indicating 0%, 40%, 16.6%, and 7% bits error respectively while mobile speed of 90km/hr gives BER values of 0%, 60%, 24.8% and 10.8% respectively. The comparisons are made from the results obtained to determine the best out of the three different fading channels. It has been confirmed that Log-normal fading channel gives the lowest BER value while Rayleigh gives the highest bits. Based on the results obtained in this paper, Rayleigh distribution which gives the highest BER is the best distribution to be used by mobile wireless communication engineer in the design of robust system before the implementation because if the system can perform very well in Rayleigh distribution it will perform better in other distributions.

REFERENCES

[1] Iskander, C.D (2011), "A MATLAB Based Object-oriented Approach to Multipath Fading Channel Simulation" Hi-Tek Multisystems White paper, Quebec, Canada.

[2] Bernard, S. (1997), Rayleigh Fading Channels in Mobile Digital Communication System, IEEE communication magazine, vol. 35, no. 7 pp.90-100.

[3] Adeyemo, Z.K. and Raji, T.I. (2010), Bit Error Rate Analysis for Wireless Links Using Adaptive Combining Diversity, Journal of Theoretical and Applied Information Technology, vol. 20, no.8,pp. 58-66.

[4] Mohamed-slim., Tang and Goldsmith (1999), An Adaptive Modulation Scheme for Simultaneous Voice and Data Transmission over Fading Channels, IEEE Journal on Selected Areas in Communications, vol.17, no.5, pp.837-847.

[5] Xuanming, D. (2006), Effect of Slow Fading and Adaptive Modulation on TCP (UDP) Performance of High-speed packet Wireless Networks, Technical Report no UCB/EECS-2006-109, University of California, Berkeley.

[6] Patzold, M. (2002), Mobile Fading Channels ,1st Edition, John Wiley and Sons, Norway.

[7] Vijay G. K. (2007), Wireless Communications and Networking. Morgan Kaufmann, United States of America.

[8] Rappaport, T. S. (2002), Wireless Communications: Principles and Practice 2nd Edition, Prentice Hall.

[9] Rappaport, T. S. (1989), Characterization of UHF Multipath Radio Channels in Factory Buildings, IEEE Transaction on Antennas and Propagation, vol.37, no.8, pp.1058-1069.

[10] Steven, W., Jeffrey, G. A. and Nihar, J. (2007), The Effect of Fading, Channel Inversion, and Threshold Scheduling on Ad Hoc Networks, IEEE Transmissions on Information Theory, vol.53, no.11 pp.4127-4149.

[11]   Parastoo, S. and Predrag, R. (2008), On Information Rate of Time-varying Fading Channels Modeled as Finite-State Markov Channels, IEEE Transactions on Communication, vol.56, no.8, pp.1268-1278.

[12]   Ramjee, P. and Hiroshima, H. (2002), Simulation and Software Radio for Mobile Communications, Communication Research Laboratory, Independent Administrative Institution, Japan.

[13]   Gregory, D. D., Theodore, S. R., and David, A. W (2002), New Analytical Models and Probability Density Functions for Fading in Wireless Communications, IEEE Transactions on Communications, vol.50, no.6, pp.1005-1012.

[14]   Hyunseung, C. (2006), Mobile Radio Propagation/Channel Coding Mobile Computing. Sungkyunkwan University retrieved March 11, 2011, from (www.google.com).

[15]   Taneda M.A., Takada J., and Araki K. (2001), The Problem of the Fading Model Selection. IEICE Transaction on Communication, vol. E84 B, no. 3, pp 660-666.

[16]   Peppas, K.P., Nistazakis and Tumbras, G.S. (2011), "An Overview of the Physical insight and the various Performance Metrics of Fading Channels in Wireless Communication Systems" Advanced Trends in Wireless Communications. Retrieved from http://www.intechopen.com/

[17]   Zhang, Y. and Fujise, M. (2006), "performance Analysis of Wireless Networks over Rayleigh Fading Channel", IEEE Transactions on Vehicular Technology, vol. 55, no. 5, pp. 1621-1632.

[18]   Charalambons, C.D, Buttitude, R.J.C., Li, X. and Zhan, J. (2008), "Modeling Wireless Fading Channels via Stochastic Differential Equations, Identification and Estimation Based in Noisy Measurements" IEEE Transaction on Wireless Communications, vol. 7,no. 2, pp. 434-439.

[19]   Tellambura, C. (2008), "Bound on the Distribution of a Sum of Correlated log-normal Random Variables and their Applications" IEEE Transaction on Communications, vol. 56, no. 8, pp.1241-1248.

[20]   Young, J.C. and Saewong, B., (2008), "Multichannel Wireless Scheduling under Limited Terminal Capacity" IEEE Transactions on Wireless Communications, vol. 7, no. 2, pp. 611-617.

[21]   Rashmi, S., Sunil, J., and Navneet, A. (2011), Performance Analysis of Different M-ary Modulation Techniques in Cellular Mobile Communication, In Multimedia Communications, A special Issue from IJCA, retrieved on 22nd may, 2011 from (www.ijcaonline.org).

[22]   Log-normal distribution (online) retrieved May 21, 2011, from (www.wikipedia.com/).

[23]   Shuwei, H. (2002),"Evaluation Techniques for TDMA Cellular Radio with Co-channel Interference". Unpublished PhD Thesis, University of Calgary, Alberta.

PARAMETRIC PERFORMANCE ANALYSIS OF PATCH ANTENNA USING EBG SUBSTRATE

Mst. Nargis Aktar[1], Muhammad Shahin Uddin[1], Monir Morshed[1],
Md. Ruhul Amin[2], and Md. Mortuza Ali[3]

[1]Department of Information and Communication Technology
Mawlana Bhashani Science and Technology University, Bangladesh
[2]Department of Electrical and Electronic Engineering
Islamic University of Technology, Dhaka, Bangladesh
[3]Department of Electrical and Electronic Engineering
Rajshahi University of Engineering and Technology, Bangladesh

E-mail: {nargismbstu@gmail.com, shahin.mbstu@gmail.com,
mm031087@gmail.com,aminr_bd@yahoo.com, and mmali.ruet@gmail.com}

ABSTRACT

In recent years, microstrip antennas become very popular because of their interesting advantages of low cost, low profile, light weight, easy fabrication and ease of analysis. Electromagnetic Band Gap (EBG) substrate is used as a part of antenna structure to improve the performance of the patch antenna. Usually, the performance of a patch antenna depends on the parameters such as Return Loss (RL), Bandwidth (BW), Gain, and Directivity. In this paper, we propose a rectangular microstrip patch antenna with EBG substrates of different EBG patch width and analysis the performance of the proposed antenna compare with a conventional patch antenna using the same physical dimension using HFSS simulator. Here we show four analyses of our designed antenna and compare with the conventional microstip patch antenna. Comparison results show that, the performances of our proposed schemes are better comparing the conventional patch antenna.

KEYWORDS

Microstrip Patch Antenna, Electromagnetic Band Gap (EBG) Substrates, Return Loss, Gain, Bandwidth and Directivity, HFSS simulator

1. INTRODUCTION

With the drastic demand of wireless communication system and their miniaturization, antenna design becomes more challenging. Recently patch antennas have been widely used. In spite of its several advantages, they suffer from drawbacks such as narrow bandwidth, low gain and excitation of surface waves [1]. To overcome these limitations of microstrip patch antennas two techniques have been used, namely micromachining and periodic structures called the electromagnetic band gap (EBG) structures [13]. However, in recent years there has been considerable effort in the EBG structure for antenna application to suppress the surface wave and overcome the limitations of the antenna. Many works have been done to enhance the performance of the patch antennas [2-11]. The EBG structure utilizes the inherent properties of dielectric materials to enhance the microstrip antenna performance. The characteristics of EBG depend on the shape, size, symmetry and the material used in their construction. There are four main parameters affecting the performance of mushroom like EBG structures such as EBG patch width, gap width, substrates thickness and substrates permittivity [14].

In this paper, we propose a rectangular patch antenna with EBG substrates of different EBG patch width, w and investigate the performance of the proposed antenna. The remainder of the paper is organized as follows: In section 2, a brief description of EBG structure. In section 3 present the conventional and proposed antenna design and configuration. In section 4 present the simulation results and discussion. The conclusion of this paper is provided in section 5.

2. RELATED WORK

EBG structures are artificially engineered periodic structures with the capability of prohibiting the propagation of electromagnetic waves in a specific frequency band regardless of incident angles and polarization states. They are realized by periodically loading a substrate material with metallic patches [17]. In [2] paper used the mushroom-like EBG structure which was integrated with a patch antenna in various configurations to observe the improvement on the properties of the antenna. Extensive simulation was carried out using CST Microwave Studio. It was observed that with the integration of the EBG structure various improvements were noted in terms of the return loss, directivity, gain and backlobe reduction. In [9] Parametric studies were performed to determine the effect of each parameter on the patch performance, and optimising them for the wide bandwidths and high gainsIn [3], a parametric analysis of the mushroom-type structure and the planar periodic one was presented. The effect of the patch size and the unit cell size on the simultaneous AMC and EBG behavior was discussed, and the influence of the substrate height and the permittivity was studied.

3. THEORY OF EBG

Recently there has been growing interest in utilizing electromagnetic band gap (EBG) structures in the electromagnetic and antenna community [7]. EBG structures can be defined as artificial periodic (or sometimes non-periodic) objects that prevent or assist the propagation of electromagnetic waves in a specified band of frequency for all incident angle and polarization state. The main advantage of EBG structure is their ability to suppress the surface wave current that reduce the antenna efficiency and radiation pattern. In mushroom like EBG structure, a band gap is observed between the frequency 7 GHz and 11 GHz. On the other hand, for the uniplanar EBG surfaces a band gap is observed the frequency from 13GHz to 14.6 GHz. In this paper, mushroom like EBG surface is used in order to design patch antenna on EBG substrates because it has a lower frequency band gap and a wider bandwidth than the uniplanar EBG surface. A two dimensional mushroom like EBG structure is shown in Figure 3.1. Design of patch antenna mushroom like EBG structures are preferable because light weight, low fabrication cost. The parametric study of mushroom like EBG structure is presented in [12-13]. The affecting parameters for the performance of the proposed antenna are EBG patch width w, gap width g, substrates thickness h and substrates permittivity ε_r that are directly depend on the operating wavelength of the patch antenna [8,18].The parameters are varying with operating wavelength as like this that the EBG patch width varies from $0.04\lambda_{12\ GHz}$ to $0.20\lambda_{12\ GHz}$, gap width varies from $0.01\lambda_{12\ GHz}$ to $0.12\lambda_{12\ GHz}$ and the substrate thickness varies from $0.01\lambda_{12\ GHz}$ to $0.09\lambda_{12\ GHz}$. Here, λ_{12} means the wavelength between medium 1 and 2 i.e. the free space and the guiding device and GHz means the wavelength respect to the GHz range frequency [14]. In this paper, mushroom like EBG structure is integrated with the proposed antenna, the study is focusing with different EBG patch width and the other parameters of the EBG structure is consider constant and analysis the performance parameter of the proposed antenna at different EBG patch width.

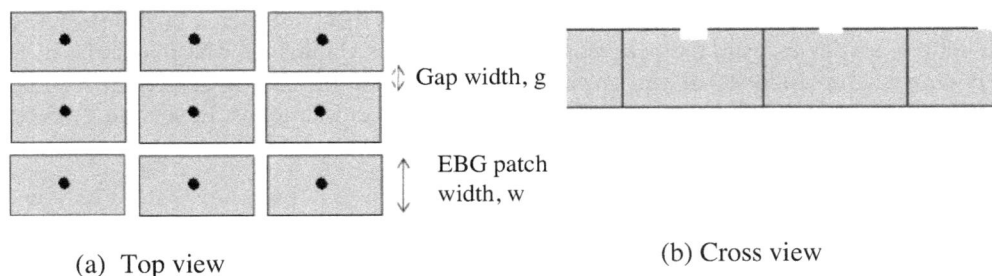

(a) Top view   (b) Cross view

Figure 3.1 Two dimensional mushrooms like EBG surfaces: (a) Top view (b) Cross view

4.   ANTENNA DESIGN AND CONFIGURATION

In order to identify and verify the enhancement of the performance of patch antenna on EBG substrates, designed a conventional antenna and the proposed antennas. The width, W of the rectangular patch antenna is usually chosen to be larger than the length of the patch, L to get higher bandwidth. The resonant frequency of the antenna is 10 GHz. In this paper, we use Neltec dielectric material as patch substrates whose dielectric constant is 2.45 that is low dielectric constant, in low dielectric constant surface wave losses are more severe and dielectric and conductor losses are less severe. By using EBG substrates, surface wave loss can be reduced easily. The antenna is excited by a microstrip transmission line feed. The point of excitation is adjustable to control the impedance match between feed and antenna, polarization, mode of operation and excitation frequency [15-16]. Table1 shows the important parameters for the geometrical configuration of the patch antenna.

Table1 Geometrical configuration of the patch antenna

Antenna Part	Parameter	Value
Patch	Length	8.8mm
	Width	11.4mm
Patch Substrates (NeltecN×9245) (IM)(tm)	Dielectric constant	2.45
	Height	0.787mm
	Dielectric loss tangent	0.01
EBG Substrates	EBG patch Width	$(0.08 \text{ to } 0.20)\lambda_{12 \text{ GHz}}$
	Gap Width	$0.02\lambda_{12 \text{ GHz}}$
	Substrates thickness	$0.04\lambda_{12 \text{ GHz}}$
Operating Frequency		10GHz

In this paper, to analysis the performance parameter of the proposed antenna at different EBG patch width w, we have designed five antennas integrated with different EBG patch width and defined their name as Antenna-1, Antenna-2, Antenna-3, Antenna-4, Antenna-5 that are given below

Antenna-1:

In case of Antenna-1, EBG patch width, $w = 0.08\lambda_{12\,GHz}$, gap width, $g = 0.02\lambda_{12\,GHz}$, substrates thickness, $h = 0.04\lambda_{12\,GHz}$

Antenna-2:

In case of Antenna-2, EBG patch width, $w = 0.10\lambda_{12\,GHz}$, gap width, $g = 0.02\lambda_{12\,GHz}$, substrates thickness, $h = 0.04\lambda_{12\,GHz}$

Antenna-3:

In case of Antenna-3, EBG patch width, $w = 0.12\lambda_{12\,GHz}$, gap width, $g = 0.02\lambda_{12\,GHz}$, substrates thickness, $h = 0.04\lambda_{12\,GHz}$

Antenna-4:

In case of Antenna-4, EBG patch width, $w = 0.16\lambda_{12\,GHz}$, gap width, $g = 0.02\lambda_{12\,GHz}$, substrates thickness, $h = 0.04\lambda_{12\,GHz}$

Antenna-5:

In case of Antenna-5, EBG patch width, $w = 0.20\lambda_{12\,GHz}$, gap width, $g = 0.02\lambda_{12\,GHz}$, substrates thickness, $h = 0.04\lambda_{12\,GHz}$

5. SIMULATION RESULTS AND DISCUSSION

Now-a-days, it is a common practice to evaluate the system performances through computer simulation before the real time implementation. A simulator "Ansoft HFSS" based on finite-element method (FEM) has been used to calculate return loss, impedance bandwidth, radiation pattern and gains. This simulator also helps to reduce the fabrication cost because only the antenna with the best performance would be fabricated.

5.1 RETURN LOSS

The simulated return losses that are obtained from conventional antenna and the antennas using EBG substrates are given in Table 5.1. From the table, it is seen that the return loss of the antenna with EBG structure is less compared to the conventional antenna. It is also seen that when the EBG patch width increases than the return loss also increases. Therefore, the antenna performance is better than the conventional antenna because the return loss is reduced for the EBG structure.

Table 2 Return loss of conventional antenna and the proposed antennas using EBG substrates

Antenna Name	EBG Patch width	Return Loss in dB
Conventional Antenna		-14.5
Antenna-1	$0.08\lambda_{12\,GHZ}$	-23
Antenna-2	$0.10\,\lambda_{12\,GHZ}$	-22.8
Antenna-3	$0.12\,\lambda_{12\,GHZ}$	-21
Antenna-4	$0.16\,\lambda_{12\,GHZ}$	-20
Antenna-5	$0.20\,\lambda_{12\,GHZ}$	-19.3

The return loss versus EBG patch width of the proposed antenna is shown in Figure 5.1

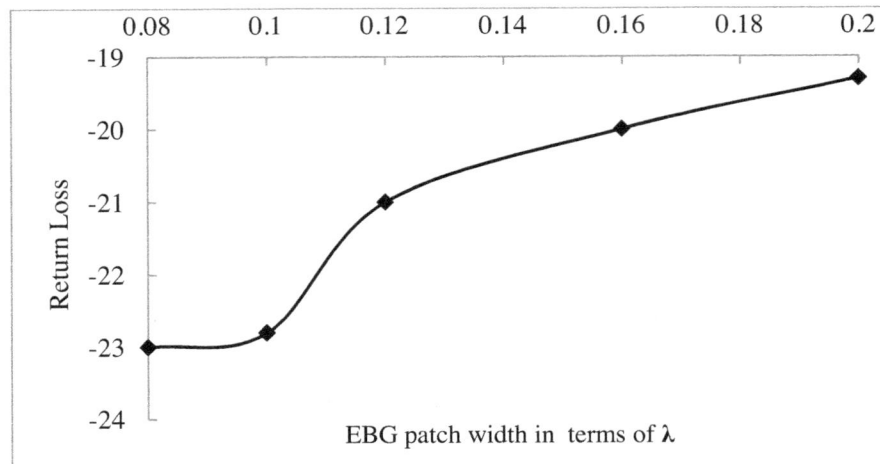

Fig 5.1 Return loss versus EBG patch width of the proposed antenna

Figure 5.1 shows the simulated results of the return loss of a patch antenna on EBG structures with different EBG patch width. Here the return loss is increasing when the EBG patch width is increasing.

5.2   BANDWIDTH (BW)

The simulated bandwidth that are obtained from the conventional antenna and the antennas integrated with EBG structures are given in Table 3

Table 3 Bandwidth of conventional antenna and the antennas on EBG substrates

Antenna Name	EBG patch width	Bandwidth in MHz
Conventional Antenna		240
Antenna-1	$0.08\lambda_{12\,GHZ}$	330
Antenna-2	$0.10\,\lambda_{12\,GHZ}$	330
Antenna-3	$0.12\,\lambda_{12\,GHZ}$	320
Antenna-4	$0.16\,\lambda_{12\,GHZ}$	300
Antenna-5	$0.20\,\lambda_{12\,GHZ}$	280

From the table 3, it is seen that the bandwidth of the antenna with EBG structure is higher than the conventional antenna. Therefore, the performance of the antenna with EBG structure is better than the conventional antenna because the bandwidth is increased for the EBG structure.
The bandwidth versus EBG patch width of the antenna using EBG substrates is shown in the Figure 5.2
From the Figure 5.2, it is seen that, firstly bandwidth is constant but when increase the EBG patch width then decrease the bandwidth.

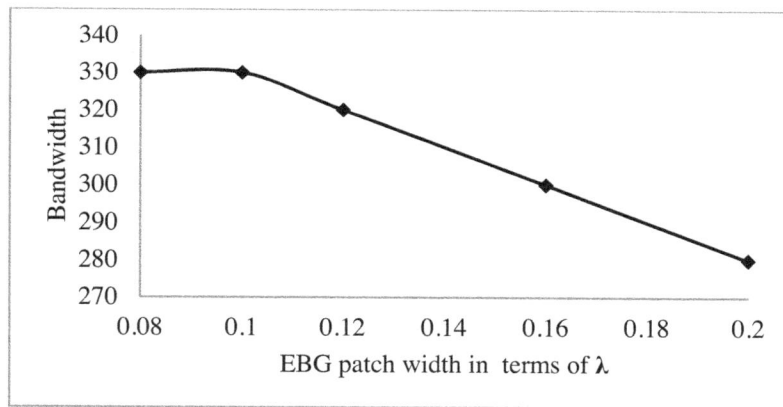

Figure 5.2 Bandwidth versus EBG patch width of the antenna on EBG substrates

5.3 GAIN

The simulated gains that are obtained from conventional antenna and the antennas using EBG substrates are in Table 4

Table 4 Gain of conventional antenna and the proposed antennas on EBG substrates

Antenna Name	EBG patch width	Gain in dBi
Conventional Antenna		22.3
Antenna-1	$0.08\lambda_{12\,GHZ}$	24.6
Antenna-2	$0.10\,\lambda_{12\,GHZ}$	25.7
Antenna-3	$0.12\,\lambda_{12\,GHZ}$	25.7
Antenna-4	$0.16\,\lambda_{12\,GHZ}$	24.4
Antenna-5	$0.20\,\lambda_{12\,GHZ}$	24.2

Table 4 shows the simulation results of the gain of conventional antenna and the antennas on EBG substrates with different EBG patch width. The table shows that the gain of the proposed antenna is higher than the conventional antenna.

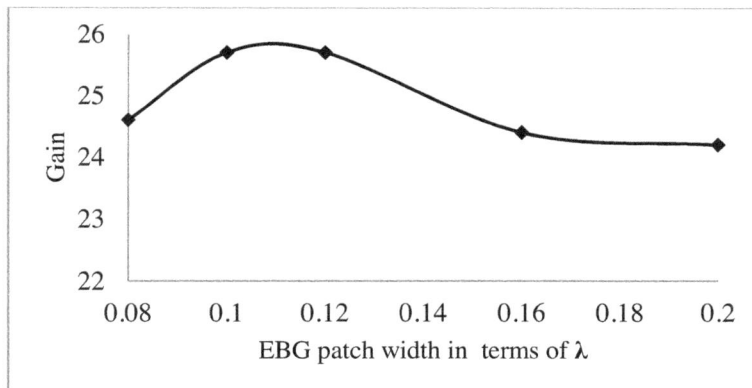

Figure 5.3 Gain versus EBG patch width of the antenna on EBG substrates

The gain versus EBG patch width of the proposed antenna is shown in the Figure 5.3. Here shows the simulated results of the gain variation of a patch antenna on EBG structures with different EBG patch width. The Figure shows that the gain is increasing with the EBG patch width from 0.08 to 0.1 and then the middle part the gain is almost constant. After that EBG patch width from 0.12, the gain is decreasing in nature with the increasing EBG patch width.

5.4   DIRECTIVITY

The simulated directivity that are obtained from conventional antenna and the antennas on EBG substrates are given in Table 5

Table 5 below shows the simulation results of the directivity of the conventional antenna and the antenna on EBG substrates with different EBG patch width. From the table, it is seen that the directivity of the antenna with EBG structure is higher than the conventional antenna.

Table 5 Directivity of conventional antenna and the antennas on EBG substrates

Antenna Name	EBG patch width	Directivity in dB
Conventional Antenna		23.4
Antenna-1	$0.08\lambda_{12\,GHZ}$	25.5
Antenna-2	$0.10\,\lambda_{12\,GHZ}$	25.6
Antenna-3	$0.12\,\lambda_{12\,GHZ}$	25.6
Antenna-4	$0.16\,\lambda_{12\,GHZ}$	24.3
Antenna-5	$0.20\,\lambda_{12\,GHZ}$	24.1

Figure 5.4 shows the simulated results of the directivity variations of a patch antenna on EBG structures with different EBG patch width. The Figure shows that the directivity is increasing with the EBG patch width from 0.08 to 0.1 and then the middle part the directivity is almost constant. After that EBG patch width from 0.12, the directivity is decreasing in nature with the increasing EBG patch width.

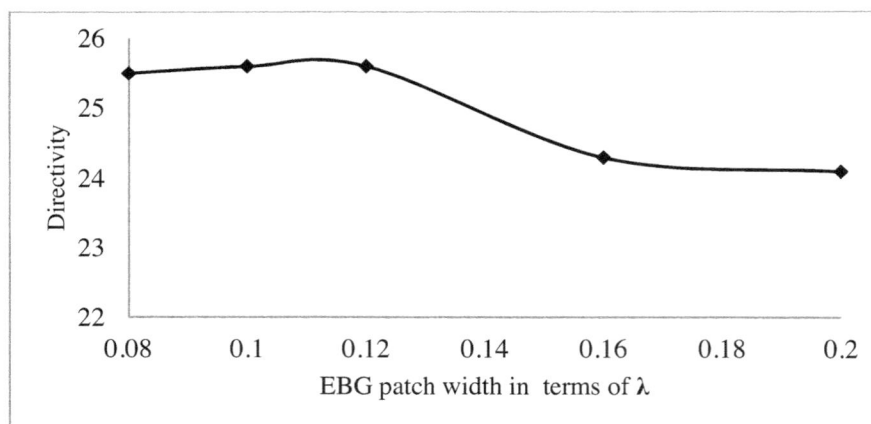

Fig 5.4 Directivity versus EBG patch width of the antenna on EBG substrate

6. CONCLUSION

The most common application of patch antenna is in mobile communication. The goals of this paper is to design conventional patch antenna and the patch antenna integrated with EBG substrates of different EBG patch width, with same physical dimensions, that can operate at 10GHz and study the performance parameter of patch antenna when EBG structure added on it. From the simulated results, it is seen that the performance is better of a patch antenna that is designed on EBG substrates than the conventional patch antenna. It is also seen from the graphical analysis of performance parameter of patch antenna with EBG substrates, the most suitable performance is obtained at the range of EBG patch width is $0.08\lambda_{12\,GHz}$ to $0.12\,\lambda_{12\,GHz}$. In future, our targets are to real time implementation of the proposed antenna and also design another patch antenna with EBG substrates that can operate at higher frequency.

REFERENCES

[1] F.Benikhlef, N. Boukli-Hacene, "Influence of Side Effect of EBG Structures on the Far-Field Pattern of Patch Antennas", IJCSI International Journal of Computer Science Issues, Vol. 9, Issue 1, No 3, January 2012.

[2] G.Gnanagurunathan and U.G.Udofia, "Performance analysis of the mushroom-like-EBG structure integrated with a microstrip patch antenna," Proceedings of IEEE Asia-Pacific Conference on Applied Electromagnetic (APACE), 2010.

[3] Peter KOVÁCS, Zbyněk RAIDA, Marta MARTÍNEZ-VÁZQUEZ "Parametric Study of Mushroom-like and Planar Periodic Structures in Terms of Simultaneous AMC and EBG Properties," Radio Engineering, Vol. 17, No. 4, December 2008.

[4] Zhixia Wei; Chenjiang Guo; Jun Ding; Nanjing Li; Zhibin Wei, "Parametric analysis of the EBG structure and its application for low profile wire antennas," 8th International Symposium on Antennas, Propagation and EM Theory(ISAPE), 2008.

[5] Jing Liang, and Hung-Yu David Yang, "Radiation Characteristics of a Microstrip Patch over an Electromagnetic Bandgap Surface," IEEE Transactions on Antennas and Propagation,Vol.55, pp1691-1697, June 2007.

[6] D. N. Elsheakh, H. A. Elsadek, E. A. Abdallah, H. Elhenawy, and M. F. Iskander, "Enhancement of Microstrip Monopole Antenna Bandwidth by Using EBG Structures," IEEE Antennas and Wireless Propagation Letters, vol. 8, pp 959-962 , 2009.

[7] Fan Yang, Member, and Yahya Rahmat-Samii," Microstrip Antennas Integrated With Electromagnetic Band-Gap (EBG) Structures: A Low Mutual Coupling Design for Array Applications," IEEE Transactions on Antennas and Propagation, Vol. 51, No. 10, October 2003.

[8] Ram´on Gonzalo, Peter de Maagt, and Mario Sorolla, "Enhanced Patch-Antenna Performance by Suppressing Surface Waves Using Photonic-Bandgap Substrates," IEEE Transactions on Microwave Theory and Techniques, Vol. 47, pp. 2131-2138, November 1999.

[9] D. Qu, L. Shafai and A. Foroozesh, "Improving Microstrip Patch Antenna Performance Using EBG Substrates," IEE Proceedings Micro Antennas Propagation, Vol.153, pp.558-563, December 2006.

[10] Atsuya Ando, Kenichi Kagoshima, Akira Kondo, and Shuji Kubota, "Novel Microstrip Antenna With Rotatable Patch Fed by Coaxial Line for Personal Handy-Phone System Units," IEEE Transactions on Antennas and Propagation, Vol. 56, pp.2747-2751, August 2008.

[11] Mr. Pramod Kumar.M , Sravan kumar, Rajeev Jyoti , VSK Reddy, PNS Rao,"Novel Structural Design for Compact and Broadband Patch Antenna," IEEE International Workshop on Antenna Technology (iWAT), 2010.

[12] R. Chantalat, C. Menudier, M. Thevenot, T. Monediere, E. Arnaud, and P. Dumon, "Enhanced EBG Resonator Antenna as Feed of a Reflector Antenna in the Ka Band," IEEE Antennas and Wireless Propagation Letters, Vol. 7 pp.349-353, 2008.

[13] Ioannis Papapolymerou, Rhonda Franklin Drayton, and Linda P. B. Katehi, "Micromachined patch Antennas," IEEE Transaction on Antenna Propagation, Vol.46, No.2, pp.275-283, 1998.

[14] Fan Yang, Yahya Rahmat Sami, "Electromagnetic band Gap Structures in Antenna Engineering," Cambridge University Press 2009.

[15] Constantine A. Balanis, "Antenna Theory Analysis and Design," Third Edition, John Wiley & Sons, 2005.

[16] D. M. Pozar, "Microwave Engineering," Third Edition New York, Wiley, 2005.

[17] Fan Yang and Yahya Rahmat-Samii, Electromagnetic Band Gap Structures in antenna engineering, Cambridge University press, 2009.

[18] Fan Yang and Yahya Rahmat-Samii, "Microstrip Antennas Integrated With Electromagnetic Band-Gap (EBG) Structures: A Low Mutual Coupling Design for Array Applications," *IEEE Transactions on Antennas and Propagation*, Vol. 51, No. 10, October 2003

Permissions

List of Contributors

Helmi Chaouech
National Engineering School of Tunis, University of El-Manar, Tunis, Tunisia
Innov'COM Laboratory, Sup'COM, University of Carthage, Tunis, Tunisia

Ridha Bouallegue
Innov'COM Laboratory, Sup'COM, University of Carthage, Tunis, Tunisia

Fawaz Bokhari
Department of Computer Science and Engineering, The University of Texas at Arlington, Texas, US

Gergely Zaruba
Department of Computer Science and Engineering, The University of Texas at Arlington, Texas, US

Martin Crew
University of Cape Town, South Africa

Osama Gamal Hassan
Cairo University, Egypt

Mohammed Juned Ahmed
King Abdullah University of Science and Technology, Saudi Arabia

Farshad Safaei
Faculty of ECE, Shahid Beheshti University G.C., Evin 1983963113, Tehran, IRAN

Hamed Mahzoon
Department of System Innovation Graduate School of Engineering Science Osaka University, Osaka, Japan

Mohammad Sadegh Talebi
School of Computer Science Institute for Research in Fundamental Sciences (IPM) Tehran, Iran

Moon Ho Lee
Division of Electronics & Information Engineering, Chonbuk National University, Jeonju 561-756, Korea

Md. Hashem Ali Khan
Division of Electronics & Information Engineering, Chonbuk National University, Jeonju 561-756, Korea

Daechul Park
Dept. of Information & Communication Engineering, Hannam University, Daedeok- Gu, Daejeon 306-791, Korea

V Ramchand
School of Computer and Systems Sciences, Jawaharlal Nehru University, New Delhi, India

D.K. Lobiyal
School of Computer and Systems Sciences, Jawaharlal Nehru University, New Delhi, India

Manish Devendra Chawhan
Assistant Prof, Dept of E&C, SRKNEC, India

Dr Avichal R. Kapur
Dean(QA) &Advisor, NYSS, MGI, India

Tony Tsang
Hong Kong Polytechnic University Hung Hom, Hong Kong

Sandhya Chilukuri
Department of Computer Engineering, M. S. Ramaiah School of Advanced Studies, Bengaluru, India

Rinki Sharma
Department of Computer Engineering, M. S. Ramaiah School of Advanced Studies, Bengaluru, India

Deepali. R. Borade
Department of Computer Engineering, M. S. Ramaiah School of Advanced Studies, Bengaluru, India

Govind R. Kadambi
Department of Computer Engineering, M. S. Ramaiah School of Advanced Studies, Bengaluru, India

Rasha S. El-Khamy
Alexandria Higher Institute of Engineering and Technology, Alexandria, Egypt

ShawkyShaaban
Dept. of Electrical Engineering, Faculty of Engineering, Alexandria Univ., Alexandria 21544, Egypt

Ibrahim Ghaleb
Dept. of Electrical Engineering, Faculty of Engineering, Alexandria Univ., Alexandria 21544, Egypt

Hassan Nadir Kheirallah
Dept. of Electrical Engineering, Faculty of Engineering, Alexandria Univ., Alexandria 21544, Egypt

Karima Ben Hamida El Abri
Syscoms Laboratory, National Engineering School of Tunis, Tunisia

Ammar Bouallegue
Syscoms Laboratory, National Engineering School of Tunis, Tunisia

Ramprasad Subramanian
Centre for Real-time Information Networks, School of Computing and Communications, Faculty of Engineering and Information Technology, University of Technology Sydney, Sydney, Australia

Kumbesan Sandrasegaran
Centre for Real-time Information Networks, School of Computing and Communications, Faculty of Engineering and Information Technology, University of Technology Sydney, Sydney, Australia

Somayeh Kafaie
School of Computer Engineering, Iran University of Science and Technology, Tehran, Iran

Omid Kashefi
School of Computer Engineering, Iran University of Science and Technology, Tehran, Iran

Mohsen Sharifi
School of Computer Engineering, Iran University of Science and Technology, Tehran, Iran

S. Taruna
Computer Science Department, Banasthali University, Rajasthan, India

Sheena Kohli
Computer Science Department, Banasthali University, Rajasthan, India

G.N.Purohit
Computer Science Department, Banasthali University, Rajasthan, India

Z.K. Adeyemo
Department of Electronic & Electrical Engineering, Ladoke Akintola University of Technology, Ogbomoso, Nigeria

D.O. Akande
Department of Electronic & Electrical Engineering, Ladoke Akintola University of Technology, Ogbomoso, Nigeria

F.K. Ojo
Department of Electronic & Electrical Engineering, Ladoke Akintola University of Technology, Ogbomoso, Nigeria

H.O. Raji
Department of Electronic & Electrical Engineering, Ladoke Akintola University of Technology, Ogbomoso, Nigeria

Mst. Nargis Aktar
Department of Information and Communication Technology Mawlana Bhashani Science and Technology University, Bangladesh

Muhammad Shahin Uddin
Department of Information and Communication Technology Mawlana Bhashani Science and Technology University, Bangladesh

Monir Morshed
Department of Information and Communication Technology Mawlana Bhashani Science and Technology University, Bangladesh

Md. Ruhul Amin
Department of Electrical and Electronic Engineering Islamic University of Technology, Dhaka, Bangladesh

Md. Mortuza Ali
Department of Electrical and Electronic Engineering Rajshahi University of Engineering and Technology, Bangladesh